TOPSELL'S HISTORIES OF BEASTS

TOPSELL'S HISTORIES OF BEASTS

MALCOLM SOUTH, Editor

Nelson-Hall nh Chicago

To Yvonne, Mark, and Elizabeth

LIBRARY OF CONGRESS CATALOGING IN PUBLICATION DATA

Topsell, Edward, 1572–1625?
   Topsell's Histories of beasts.

   Consists of selections from the author's The history of four-footed beasts and The history of serpents.
   Includes bibliographical references.
   1. Zoology—Pre-Linnean works. I. South, Malcolm, 1937– . II. Topsell, Edward, 1572–1625? Historie of foure-footed beastes. Selections.
III. Topsell, Edward, 1572–1625?. Historie of serpents. Selections. IV. Title. V. Title: Histories of beasts.
QL41.T67   1981   590   80-28838
ISBN 0-88229-642-6

Copyright © 1981 by Malcolm South

All rights reserved. No part of this book may be reproduced in any form without permission in writing from the publisher, except by a reviewer who wishes to quote brief passages in connection with a review written for broadcast or for inclusion in a magazine or newspaper. For information address Nelson-Hall Inc., Publishers, 111 North Canal Street, Chicago, Illinois 60606.

Manufactured in the United States of America

10 9 8 7 6 5 4 3 2 1

CONTENTS

LIST OF ILLUSTRATIONS — vii
ACKNOWLEDGMENTS — ix
INTRODUCTION, BY MALCOLM SOUTH — xi

THE EPISTLE DEDICATORY — 1
THE BEASTS — 7
  THE APE — 9
  THE ASP — 15
  THE BEAR — 21
  THE CAMEL — 27
  THE CAMELOPARDAL — 33
  THE CAT — 37
  THE COCKATRICE — 43
  THE CROCODILE — 49
  THE DOG — 61
  THE DRAGON — 75
  THE ELEPHANT — 87
  THE FOX — 99
  THE GULON — 107
  THE HYENA — 111
  THE LAMIA — 119
  THE LION — 125
  THE MANTICHORA — 143
  THE PANTHER, COMMONLY CALLED A PARDAL,
    A LEOPARD, AND A LIBBARD — 147
  THE RHINOCEROS — 157
  THE TIGER — 163
  THE UNICORN — 169
  THE WOLF — 175
AN EPILOGUE TO THE READERS — 185

LIST OF ILLUSTRATIONS

THE APE
8

ASPS
14

THE BEAR
20

THE BACTRIAN CAMEL
26

THE TWO SORTS OF CAMELOPARDALS
32

THE CAT
36

THE COCKATRICE
42

NILE CROCODILE
48

A GROUP OF HOUNDS AND A TYPE
OF SPANIEL OR POODLE
60

TWO TYPES OF WINGED DRAGONS
74

THE ELEPHANT
86

THE FOX
98

THE GLUTTON OR "GULON,"
THE EUROPEAN WOLVERINE
106

The Hyena
110

The Lamia
118

The Lion
124

The Manticore or Mantichora
142

The Panther or Leopard,
or "Pardal, Libbard"
146

The Rhinoceros
156

The Tiger
162

The Unicorn
168

The Wolf
174

ACKNOWLEDGMENTS

I AM GRATEFUL to Muriel Bagby, William Bloodworth, and Robert McCutcheon for their help. I have been helped in many ways by the reference staff of Joyner Library at East Carolina University. My special appreciation goes to my wife, who typed the manuscript. A grant from East Carolina University provided part of the financial support for the book.

I have used a second edition of Topsell found in the Rare Book Room of Duke University Library. I wish to thank the staff for their courtesy and help. The following illustrations in my book have been reproduced with the permission of Duke University Library: Asps; the Cockatrice; and the Lamia.

The following illustrations have been reproduced with the permission of Dover Publications, Inc., from *Curious Woodcuts of Fanciful and Real Beasts,* with picture selection and layout by Edmund V. Gillon, Jr. (New York: Dover Publications, Inc., 1971): the Ape (p. 8); the Bear (p. 20); the Bactrian Camel (p. 26); the Two Sorts of Camelopardals (p. 32); the Cat (p. 36); Nile Crocodile (p. 48); a Group of Hounds and a Type of Spaniel, or Poodle (p. 60); Two Types of Winged Dragons (p. 74); the Elephant (p. 86); the Fox (p. 98); the Glutton, or "Gulon," the European Wolverine (p. 106); the Hyena (p. 110); the Lion (p. 124); the Manticore, or Mantichora (p. 142); the Panther, or Leopard, or "Pardal, Libbard" (p. 146); the Rhinoceros (p. 156); the Tiger (p. 162); the Unicorn (p. 168); and the Wolf (p. 174).

INTRODUCTION

BY MALCOLM SOUTH

THIS BOOK CONSISTS of selections from two Renaissance natural histories by Edward Topsell: *The History of Four-Footed Beasts* (1607) and *The History of Serpents* (1608). Topsell was born in 1572; he took his B.A. and M.A. from Cambridge and became a minister of the Church of England. During his career, he held several livings and wrote three books besides *The History of Four-Footed Beasts* and *The History of Serpents*. It appears he died in 1638.[1] His two natural histories, along with a work called *The Theatre of Insects*, were reissued in one volume in 1658.

During Topsell's time, there was a body of traditional animal lore that had been passed down through the centuries. Renaissance writers on natural history drew upon recognized authorities, and preeminent among them were five ancient writers: Aristotle (384–322 B.C.); Pliny the Elder (A.D. 23–79); Oppian (c. A.D. 180); Gaius Julius Solinus (c. A.D. 218); and Claudius Aelianus (c. A.D. 220).[2] The greatest authority on zoology in the Renaissance was Swiss naturalist Konrad Gesner, the author of *Historia Animalium* (five volumes, 1551–1587). Gesner drew upon ancient writers as well as modern writers. Topsell was a transcriber of Gesner. *The History of Four-Footed Beasts* was largely a translation of Gesner, and much of *The History of Serpents* was based on Gesner.

Both Gesner and Topsell sought to separate truth from fiction; but because zoology was not far advanced and much fabulous lore about animals was generally accepted, they had difficulty distinguishing the real from the fabulous. One aim of both writers was to give a detailed "history" of each beast, and this involved including strange beliefs and fables about it. Topsell often cautions his readers that he is including a belief on the authority of another writer and that they should use their judgment about its truthfulness.

Topsell's accounts of beasts have a special charm, because he shows personal feelings and gives each beast a "character" of its own. For example, he describes the hyena as an ignoble, hypocritical animal. In contrast, he sees the elephant as a noble beast, whose great size and strength are a demonstration of God's power. In Topsell's eyes, there is an eternal battle between good and evil; and he repeatedly draws moral lessons from the animal world. For instance, the gulon is said to be an insatiable glutton: it feeds on carcasses; and, when it finds one, it eats as much as it can and then disgorges itself by drawing its body through a narrow passage between two trees. Then it gorges itself again and empties its belly in the same way; thus it continues gorging and disgorging itself until it has devoured the entire carcass. Topsell uses the gulon to moralize about the evils of gluttony and drunkenness. He condemns people who offer "sacrifice to nothing but their bellies" and thereby "lose the marks of humanity, reason, memory, and sense."

In the editions of 1607 and 1608, Topsell's two volumes comprise about eleven hundred folio pages. There are approximately one hundred thirty general headings for four-footed beasts and serpents, and some general headings have subdivisions. Because of space considerations, I have limited my book to twenty-two animals, and I have omitted or reworked some material in each of these twenty-two histories. Eighteen of the animals come from *The History of Four-Footed Beasts;* the other four (the asp, the cockatrice, the crocodile, and the dragon) come from *The History of Serpents.* The animals I have selected are ones that should appeal to most readers. A number of animal families are represented, and there are both domesticated and wild animals. Several histories deal with fabulous animals, and these histories are among the most interesting ones in Topsell.

The matter of length is not the only reason for omissions and revisions. This book is not designed for the specialist but for the general reader, and in those places where I have felt that I could gain readability, I have abridged or reworked material. I have revised confusing sentences and passages and omitted sections of limited interest. With the exception of some words where no modern equivalents could be found, I have used modern spelling. At the same time, I have sought to preserve the tone of the original work as much as possible. For example, I usually keep archaic words if their meanings are clear from the context, and, unless confusion exists, I do not change Topsell's sentence structure and grammar.

I have listed the beasts in alphabetical order. (This is the order that Topsell has used.) I have used "The Epistle Dedicatory" to *The History of Four-Footed Beasts* as an introduction to the beasts, and "An Epilogue to the Readers" from the same work serves as a conclusion. I have written a note to accompany each history and to give some basic information about each animal. In each case the note precedes the history. Topsell's volumes contain many woodcut illustrations; although I was unable to include all of the illustrations found in the histories I have chosen, I have included at least one illustration in each history.[3]

---

[1]*Dictionary of National Biography*, s. v. "Topsell, Edward"; and M. St. Clare Byrne, *The Elizabethan Zoo* (London: F. Etchells & H. Macdonald, 1926), pp. ix—xi.

[2]For a good discussion of the sources of animal lore, see P. Ansell Robin, *Animal Lore in English Literature* (London: John Murray, 1932), pp. 3—20.

[3]For information about the woodcuts in Gesner and Topsell, see "Publisher's Note," *Curious Woodcuts of Fanciful and Real Beasts* (New York: Dover Publications, 1971), pp. v—vii. This book contains excellent reproductions of 190 woodcuts from the works of Gesner and Topsell. Most of the illustrations in this book are reproduced from the Dover edition.

THE EPISTLE DEDICATORY

THE EPISTLE DEDICATORY

To the Reverend and Right Worshipful
Richard Neile, Doctor of Divinity,
Dean of Westminster, Master of the Savoy, and
Clerk of the King His Most Excellent Majesty's Closet;
All Felicity Temporal, Spiritual, and Eternal

RIGHT WORTHY AND learned Dean, my most respected patron, I shall endeavor to prove unto you that this work which I now publish under the patronage of your name is divine and necessary for all men to know, that the work is true and without slander and suspicious scandal to be received, and that no man ought to publish this unto the world other than a divine or preacher.

I see no cause why any man should doubt that the knowledge of the beasts, like the knowledge of the other creatures and works of God, is divine, seeing that at the first they were created and brought to man, and all by the Lord Himself. So their life and creation are divine in respect to their maker, and their naming was divine in respect to Adam, for out of the plenty of his own divine wisdom he gave them their names, as it were out of a fountain of prophecy, foreshadowing in one elegant and significant denomination the nature of every kind. When I affirm that the knowledge of beasts is divine, I mean no other than the right and perfect description of their names, figures, and natures. This is in the Creator divine, and therefore such as is the fountain, such are the streams issuing from the same into the minds of men.

In Holy Scripture there were of beasts three holy uses. The first use was for sacrifices. The next use was in visions, as the visions of Daniel, Ezekiel, and St. John show. The third and last holy use was for reproof and instruction to man.

Of the use for reproof and instruction we find sundry examples which we ought to heed. Who is so unnatural and unthankful to his parents but by reading how the young storks and woodpeckers do in their parents' old age feed and nourish them, will not repent, amend his folly, and be more natural? Where is there such a sluggard and drone that, in considering the labors, pains, and travels of the emmet, little bee, field mouse, squirrel, and other such, will not learn for shame to be more industrious and set his fingers to work? Why should any man living fall to do evil against his conscience or at the temptation of the Devil seeing that a lion will never yield? And what king is not invited to clemency and dissuaded from tyranny seeing that the king of bees has a sting but never uses it? How great are the love and faithfulness of dogs, the meekness of elephants, the neatness and politure of the cat and peacock, the care of the nightingale to make her voice pleasant, the chastity of the turtledove, the canonical voice and the watchfulness of the cock, and the utility of a sheep.

That science is divine and ought to be inquired and sought after where the knowledge of God, the knowledge of man, the precepts of virtue, and the means to avoid evil are to be learned.

The second thing which I have promised to affirm in this discourse is the truth of the history of the creatures, for the mark of a good writer is to follow truth and not deceivable fables. I would not have the reader of these histories imagine that I have inserted or related all that is ever said of these beasts but only so much as is said by many. And if, at any time, I have set down only a single testimony, it was because the matter was clear and needed no further proof, or else I have laid it upon the author with special words, not giving the reader any warrant from me to believe it.

I have taken regard to imitate the best writers, which was easy for me to do, because Gesner relates every man's opinion; and if at any time he seemed obscure, I turned to the books which I had at hand to guess their meaning, putting in that which he had left out of many good authors and leaving out many magical devices. Although I have used no small diligence or care in collecting those things which are most essential to every beast and are most true without exception and most evident by the testimony of many good authors, yet I have delivered in this treatise many strange and rare things not as fictions but as miracles of nature for wise men to behold and observe to their singular comfort if they love the power, glory, and praise of their Maker, not withholding their consent to the things expressed because they deal with living things made by God Himself.

Now I do in a sort challenge a consent unto the probability of these things to wise and learned men, although no belief. For consent is a cleaving or yielding to a relation until the manifestation of another truth. When any man shall justly reprove anything I have written for false and erroneous, I will not stick to release the reader's consent but make satisfaction for usurpation. I care not for the rude and vulgar sort (who being utterly ignorant of the operation of learning do immediately condemn all strange things which are not engraven in the palms of their own hands or evident in their own herds or flocks). I care not, for my ears have heard some of them speak against the history of Sampson, where he tied firebrands to the tails of foxes, and many of them against the miracles of Christ. There is a tale of an old mass-priest, in the days of Henry the Eighth, who was reading in English after the translation of the Bible the miracles of the five loaves and two fishes; and, when he came to the verse that reckons the number of the guests or eaters of the banquet, he paused a little and at last said they were about five hundred. The clerk, who was a little wiser, whispered into the priest's ears that it was *five thousand*. But the priest turned back and replied with indignation, "Hold your peace, sirrah, we shall never make them believe they were five hundred."

Such priests and such people I shall draw upon my back, and, although I do not challenge a power of not erring, yet because I speak of the power of God that is unlimitable, I will be bold to aver that for truth which is not contrary to the Scriptures.

Finally, I think that no wise man will make question that it is the proper office of a preacher or divine (such being my calling) to set forth these works of God. It is my endeavor to profit and delight the reader in this book, whereinto he may look on the holiest days (not omitting prayer and the public service of God) and pass away the Sabbaths in heavenly meditations upon earthly creatures.

I have followed Doctor Gesner as near as I could. I do profess him my author in most of my stories, yet I have gathered up that which he let fall and have added many pictures and stories. In the names of the beasts and the physic I have not swerved from him at all. He was a Protestant physician (a rare thing to find any religion in a physician, although St. Luke, a physician, was a writer of the Gospel). The praises of Doctor Gesner therefore shall remain, and all living creatures shall witness for him at the last day.

Right worthy and learned Dean, this my labor whatsoever it be, I consecrate to the benefit of all our English nation under your name and patronage, a public professor, a learned and reverend divine, a famous preacher, observed in court and country. If you will vouchsafe to allow of my labors, I stand not upon others; and if this has your commendation, it shall encourage me to proceed to the rest of my labors. Therein I fear no impediment but the ability to carry out the charge, my case standing that I have not any access of maintenance but by voluntary benevolence for personal pains, receiving no more but laborious wages. And but for you, that would have also been taken from me.

I conclude with the words of St. Gregory to Leontius: "And, learning more of the good things which are declared many times about you, may we be able to entreat earnestly the omnipotent Lord for your preservation, in view of your glory."

Your Chaplain in the Church of
St. Botolph, Aldersgate

Edward Topsell

THE BEASTS

The Ape

THE APE

IN TOPSELL'S TIME, the word ape was a generic term for animals of the monkey tribe. Topsell says that apes are made for the sport and amusement of man, describes their ability to imitate human actions, and includes several of the traditional beliefs about apes. One of the best-known of these beliefs was that a mother ape with twins loved one and hated the other and that she hugged the loved one to death. A humorous tone that seems deliberate on Topsell's part appears in places in this history.

The ape is generally held for a subtle, dissembling, ridiculous, and unprofitable beast whose flesh is not good for meat nor his back for burden. The Grecians termed the ape *gelotopoios* ("made for laughter").

The philosopher Anacharsis, being at a banquet wherein divers jesters were brought in to make them laugh, never laughed. At length was brought in an ape, at the sight whereof he laughed heartily; and being demanded the cause why he had not laughed before, he answered that men do but feign merriments, whereas apes are naturally made for that purpose.

Apes are much given to imitation. They have been taught to leap, to sing, and to drive wagons, reining and whipping the horses. They are very capable of all human actions and have an excellent memory, either to show love to their friends or hateful revenge to them that have harmed them. They delight much in the company of dogs and young children, yet they will strangle young children if the children are not well looked unto.

A certain ape seeing a woman washing her child in a basin of warm water observed her diligently; and, getting into the house when the nurse was gone, it took the child out of the cradle and set water on the fire, and when the water was hot, it stripped the child naked and washed it until it killed it.

The countries where apes are found are Lybia and all the desert woods between Egypt, Ethiopia, and Lybia, and that part of Caucasus which reaches to the Red Sea. In India, they are most abundant: red, black, green, dust-colored, and white ones, all of which they bring into their cities (except the red ones, which are so lustful that they will ravish women) and present to their kings. The apes in India grow so tame that they go up and down the streets boldly and civilly as if they were children. So many apes in India showed themselves to Alexander the Great standing upright that he deemed them at first to be an army of enemies and commanded to join battle with them, until he was certified by Taxiles, a king of that country then in his camp, that they were but apes.

In Caucasus, there are trees of pepper and spices whereof apes are the gatherers. The inhabitants come and under the trees make plain a plot of ground, and afterwards they cast thereupon boughs and branches of pepper and other fruits, as it were carelessly. The apes secretly observe this, and in the night season they gather together in great abundance all the branches loaded with pepper and lay them in heaps upon that plot of ground, and so in the morning come the Indians and gather the pepper from those boughs in great measure, reaping no small advantage by the labor of the apes, who gather the fruits for them while they sleep.

In the region of Basman, subject to the great Cham of Tartaria, are many and diverse sorts of apes very like mankind. When hunters take them, they pull off all the hair except from the beard and the hole behind and afterward dry them with hot spices; and, powdering them, they sell them to merchants, who carry them about the world, persuading simple people that there are men of no greater stature.

In the land of the Troglodytae, there are apes which are maned about the neck like lions, as big as great bellwethers.

There are divers kinds of apes. Some are called *cercopitheci, monkeys, chaerogitheci, hog apes, cepi, callitriches, marmosets, cynocephali, satyrs,* and *sphinxes*. They are not all alike. Some resemble men one way, and some another.

Apes do outwardly resemble men very much. In their face, nostrils, ears, eyelids, breasts, arms, thumbs, fingers, and nails, they agree very much. Their hair is harsh and short; they are hairy in the upper part like men, and in the nether part like beasts. They have no navel, but there is a hard thing in that place. The nails of apes are half round; and when apes are in copulation, they bend their elbows before them, the sinews of their hinder joints being turned clean about; but with a man it is clean otherwise. They cannot exactly stand upright, and therefore they run and stand like a man that counterfeits a lame man's halting.

And, as the body of an ape is ridiculous by reason of an indecent likeness and imitation of man, so is his soul or spirit. For they are kept only in rich men's houses to sport withal, being for that cause easily tamed, following every action they see done even to their own harm without discretion.

There is a story that a certain ape swimming to land after a shipwreck was seen by a countryman, and, thinking the ape to be a man in the water, the man gave him his hand to save him. Yet, in the meantime, the man asked the ape what countryman he was.

When the ape answered he was an Athenian, the man asked whether he knew Piraeus (which is the port of Athens).

"Very well," said the ape, "and his wife, friends, and children."

Whereat the man did what he could to drown him.

Apes keep for the most part in caves and hollow places of hills, in rocks and trees, feeding upon apples and nuts; but if they find any bitterness in the shell, they cast all away. They eat lice and pick them out of heads and garments. They will drink wine till they are drunk, but if they drink it oft, they grow not great.

They are taken by laying for them shoes and other things. If hunters lay shoes, they are leaden ones, too heavy for the ape to wear, wherein are made such devices of gins that, when once the ape has put the shoes on, they cannot be gotten off without the help of man. Sometimes hunters catch apes in this manner: the hunters anoint their eyes with water in the presence of apes, and so departing, leave a pot of lime or honey instead of water; the ape comes and anoints its eyes therewith and so being not able to see, is taken.

They bring forth young ones for the most part by twins, and they love the one and hate the other. That which the mother loves she bears in her arms; the other one hangs at her back. And for the most part, she kills the loved one by pressing it too hard; afterward she sets her whole delight upon the other.

When the Egyptians describe a father leaving his inheritance to a son that he loves not, they picture an ape with her young one upon her back.

Apes (the kind with tails) are heavy and sorrowful when the moon is in the wane, but they leap and rejoice at the change.

Apes love conies very tenderly. In England, an old ape (scarce able to go) did defend tame conies (rabbits) against a weasel.

They fear a shellfish and a snail very greatly. In Rome, a certain boy put a snail in his hat and came to an ape, who as he was accustomed leaped upon his shoulder and took off his hat to kill lice in his head; but, espying the snail, it was a wonder to see with what haste the ape leaped from the boy's shoulder and in trembling manner looked back to see if the snail followed him. When a snail was tied to one end of the chain of another ape, so that he could not choose but continually to look upon it, one cannot imagine how the ape was tormented; finding no means to get from it, the ape cast up whatsoever was in his stomach and fell into a grievous fever till he was removed from the snail and was refreshed with wine and water.

It was an ancient custom in former times when a parricide was executed to put him (after he was whipped with bloody stripes) into a sack with a live serpent, a dog, an ape, and a cock. By the serpent was signified his extreme malice to mankind in killing his father; by the ape, that in the likeness of man he was a beast; by the dog, how like a dog he spared none, not even his own father; and by the cock, his hateful pride. Then were they all together hurled headlong into the sea, that he might be deemed unworthy of all the elements of life and other blessings of nature.

A lion rules the beasts of the earth; and a dolphin, the beasts of the sea. When the dolphin is in age and sickness, it recovers by eating a sea-ape; and so does the lion by eating an ape. Therefore, the Egyptians paint a lion eating an ape to signify a sick man curing himself.

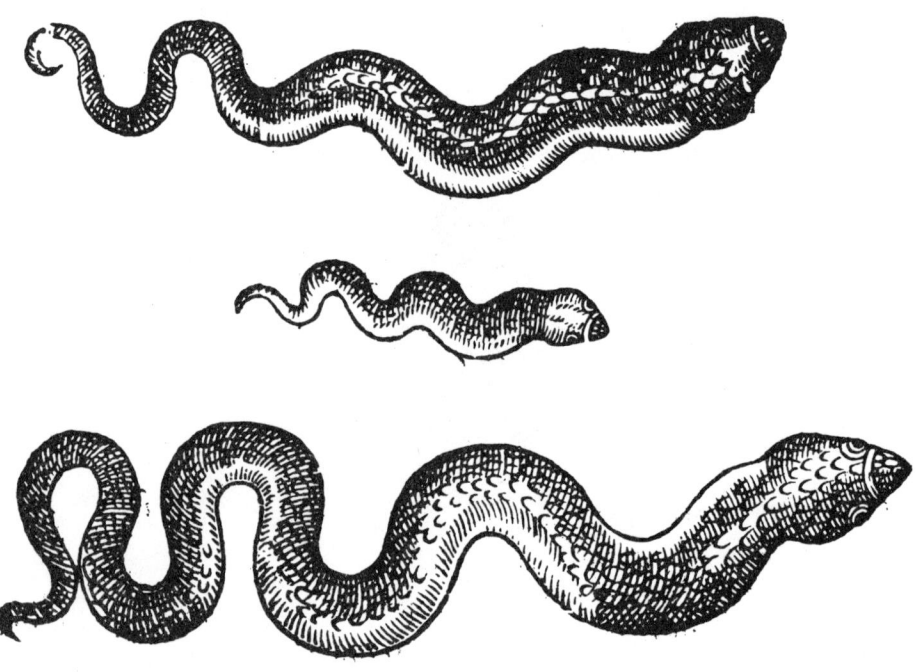

Asps

THE ASP

TOPSELL SAYS THAT *there are many kinds of asps, and organizes all the kinds in three groups: the ptyas, which poisons people's eyes by spitting out venom; the chersea, which lives on the land; and the chelidonia, which lives in riverbanks. The history describes the religious significance of the asp to the Egyptians, and there is an interesting account of the effects of the snake's poison.*

It is said that the kings of Egypt did wear the pictures of asps in their crowns whereby they signified the invincible power of principality in this creature, whose wounds cannot easily be cured. And the priests of Egypt and Ethiopia did likewise wear very long caps having toward their top a thing like a navel about which were the forms of winding asps to signify to the people that those which resist God and kings shall perish by irresistible violence. But let us leave this discourse of moralities and come nearer to the natural description of asps.

There are many kinds of asps. One kind is the dry asp. This is the longest of all other kinds, and it has eyes flaming like fire or burning coals. Another kind is called *asilus,* which does not only kill by biting but also with spitting, which it sends forth while it sets its teeth hard together and lifts up its head. Another kind is called *hirundo,* because of the similitude to swallows, for on the back it is black and on the belly white, like a swallow. There is a kind of asp called *hypnale.* It kills by sleeping, for after the wound is given, the person falls into a deep and sweet sleep wherein he dies. It has been said that this kind of asp was the kind that Cleopatra bought to bring upon herself a sweet and easy death.

I believe that all the kinds may well be reduced to three: that is, the *ptyas, chersea,* and *chelidonia.* The *ptyas* hurts by poisoning men's eyes by spitting forth venom. The *chersea* lives on land, and the *chelidonia* in riverbanks.

The asp is a small serpent like a land snake but yet of a broader back. The necks of asps swell above measure, and if they hurt while in that passion, there can be no remedy. There are two pieces of flesh like a hard skin growing out of their foreheads. Their teeth are exceedingly long and grow out of their mouths like a boar's, and through two of the longest teeth are little hollows, out of which the poison is released. The color of asps is various and divers.

The asp goes slowly, always being sleepy and drowsy. Her sight is weak, but she has a quick sense of hearing, whereby she is warned and advertised of all noise. When she hears something, she immediately gathers herself round into a circle and lifts up her terrible head. The dullness of this serpent's sight and the slowness of her pace keep her from many mischiefs and are evidence of the gracious providence of Almighty God, who has given as many remedies against evil as there are evils in the world.

The countries that breed asps are not only the regions of Africa and the confines of the Nile, but also in the northern parts of the world are many asps found. In Spain also there are asps, but none in France.

According to some writers, the Egyptians lived familiarly with asps and with continued kindness won them to be tame. They worshiped asps even as household gods, by means whereof the subtle serpents grew to a sensible conceit of their own honor and freedom and would go up and down and play with their children, doing no harm unless they were wronged. They would come and lick food from the table when they were called by a certain significant noise made by snapping the fingers. After their dinner, the guests would mix together honey, wine, and meal and then give the sign, at the hearing whereof the asps would all of them come forth from their holes; and, creeping up and lifting their heads to the table and leaving their lower parts on the ground, they licked the prepared food with great temperance, little by little, without any ravening, and then afterward departed when they were filled. And so great is the reverence that the Egyptians bear to asps that if any person has need to rise in the nighttime out of his bed, he first of all gives out the sign or token, lest he should harm the asps and so provoke them against him. When they hear the sign, all of them get them to their holes and lodgings till the person stirring is again in his bed.

The holy kind of asps they call *thermusis,* and this is used and fed in all their temples of Isis with the fat of oxen or kine. They say that this kind is not an enemy to men, except to such as are very evil. It is death to kill one of them willingly.

This kind of asp they also say is immortal and never dies; and besides, it is a revenger of sacrilege, as may appear by the following story. There was a certain Indian peacock sent to the King of Egypt; and, for its goodly proportion and form, he, out of his devotion, consecrated it to

Jupiter, and it was kept in the temple. Now, there was a certain young man who set more by his belly than his god, and he fell into a great longing to eat the said peacock; and, therefore to attain his appetite, he bribed one of the officers of the temple with a good sum of money to steal the peacock and bring it to him alive or dead. Enraged with the desire for the money, the officer sought an opportunity to steal away the peacock, and one day came to the place where he thought and knew it was kept. But when he came, he saw nothing but an asp in the place thereof; and so, in great fear, he leaped back to save his life and afterward disclosed the whole matter.

The domestical asps understand right and wrong. There is a story of such an asp, which was a female and had young ones. In her absence, one of her young ones killed a child in the house. When the old one came again, according to her custom, to seek her food, the killed child was laid forth, and so she understood the harm. Then went she and killed that young one and never more appeared in the house. It is also reported that there was a female asp that fell in love with a little boy who kept geese in the province of Egypt called Herculia, and her love to the boy was so fervent that the male of the said asp grew jealous. One day, as the boy lay asleep, he set upon him to kill him; but she, seeing the danger, awoke her love and delivered him.

Asps sort themselves by couples and live as though in marriage, male and female, so that their sense, affection, and compassion are one and the same. If it happens that one of them is killed, the other one follows the killer eagerly and will find him out even in the midst of many of his fellows. If the killer is a beast, the asp will know him among beasts of the same kind; and if the killer is a man, the asp will also find him out among men. And if the asp is let alone, he will not among thousands harm any but the killer.

There is not more mortal hatred or deadly war between any than between the ichneumon (a small weasel-like animal) and the asp. When the ichneumon has espied an asp, she first goes and calls her fellows to help her. Then, before they fight, they all do wallow in slime, or wet themselves and then wallow in the sand, as though arming their skins against the teeth of their enemy; and so, when they find themselves strong enough, they set upon the asp, bristling up their tails first of all, and they turn to the serpent until she bites at them; and then, before she can recover, with singular celerity, they fly suddenly to her chaps and tear her in pieces. The victory of this combat lies in anticipation, for if the asp first bites the ichneumon, then is she overcome; but if the ichneumon first lays hold on the asp, then is the asp overcome.

Asps bite but do not sting. When an asp has bitten a person, it is a very difficult thing to espy the place bitten or wounded, even with most excellent eyes. The reason for this is that the poison is very sharp and penetrates suddenly and forcibly under the skin, even to the inmost parts, not staying outwardly or making any great visible external appearance. The pricks of the asp's teeth are in appearance not much greater than the prickings of a needle; there is no swelling; and very little blood issues forth, and that is black in color.

After a man has been bitten by an asp, his eyes straightway grow dark and heavy, and a manifold pain arises all over his body, yet such as is mixed with some sense of pleasure. His color is all changed and appears greenish like grass. His face or forehead is bent continually with frowning, and his eyes or eyelids move up and down in drowsiness without sense.

The true signs of the biting of an asp then are stupor or astonishment, heaviness of the head, slothfulness, wrinkling of the forehead, often gaping and gnawing, nodding, bending the neck, and convulsion. Those who are hurt by the *ptyas* have blindness, pain at the heart, deafness, and swelling of the face.

So great is the effect of the poison of asps that it is worthily accounted the greatest venom and most dangerous of all other. In Alexandria, when they would put a man to a sudden death, they would set an asp to his bosom or breast and then, after the wound or biting, bid the party walk up and down; and so immediately, within two or three turns, he would fall down dead.

Some have written that a person bitten by an asp cannot be cured, but I shall show the contrary. First, it is necessary when a man is stung or bitten by a serpent that the wounded part be cut off by some skillful surgeon, or else that the flesh round about the wound, with the wound itself, be circumcised and cut with a sharp razor. Then let the hottest things be applied, even the searing iron, to the very bone. Also, before the ejection, the wound must be drawn with a cupping glass or a reed or with the naked rump of a ringdove or cock. (I mean that the very hole of the bird must be set upon the bitten place.) And because the wound is very narrow and small, it must be opened and made wider, and the blood must be drawn forth by scarifications, and then must be applied such medicinal herbs as are most opposite to poison, as rue and suchlike. Because the poison of asps congeals the blood in the veins, therefore must be applied all hot things made thin, as mithridatum and triacle dissolved in aqua vitae, and the same also dissolved into the wound. Then must the patient be accustomed to bathings, rubbing, and walking, and suchlike exercises. But, when once the wound begins to be purple, green, or black, it is a sign both of the extinguishment of the venom and also of the suffocating of natural heat. Then is nothing more safe than to cut off the member if the party is able to bear it. After cupping glasses and scarifications, there is nothing that can be more profitably applied than centory, myrrh, and opium, or sorrel after the manner of a plaster. But the body must be kept in daily motion and agitation, the wounds themselves often searched and pressed, and seawater used for fomentation.

We may also relate medicinal cures, especially of such things as are compound and received inwardly. First, after the wound, it is good to make the party vomit, and then afterward make him drink juice of yew and triacle, or in the default thereof, wine (as much of the juice as a groat weight, or rather more). For the trial of the party's recovery, give him the powder of centory in wine to drink; and if he keeps the medicine, he will live; but if he vomits or casts it up, he will die.

The Northern shepherds do drink garlic and stale ale against the bitings of asps. Others use hartwort, apium seed, and wine. The fruit of balsam, with a little powder of gentian in wine, or the juice of mints, keeps the stomach from the cramp after a man is bitten by an asp.

There is a story of two thieves who were condemned to be cast to serpents to be destroyed. Now, the morning before they came forth, they had been given citrons to eat; when they were brought to the place of execution, there were asps put forth unto them, who bit them and yet did not harm them. The next day, the reason being suspected, the prince commanded to give one of them a citron and the other none. So when they were brought forth again, the asps fell on them and slew the one who had not eaten a citron, but the other had no harm at all.

There is a proverb of one asp borrowing poison of another, or of the asp borrowing poison from the viper. This proverb has especial use when one bad man is helped or counseled by another bad man. When Diogenes saw a company of women talking together, he said merrily unto them, "The asp borrows venom of the viper." And with this, I conclude the history of the asp.

The Bear

THE BEAR

TOPSELL DESCRIBES THE BEAR *as a fierce creature with an unpredictable nature. He tells of the hibernation, or winter sleep, of bears and of methods of capturing them. Topsell is correct in saying that a tamed bear can never be trusted completely. He refutes the ancient belief that bear cubs are born without form and are licked into shape by the mother.*

There are in general two kinds of bears: a greater and a lesser. And these lesser are more apt to climb trees than the other, and neither do the lesser ever grow to so great a stature as the other.

There are bears which live on both the land and in the sea, hunting and catching fish like an otter or beaver, and these are white-colored. In the ocean islands towards the North, there are bears of a great stature, fierce and cruel, who with their forefeet do break up the hardest congealed ice on the sea or other great waters and draw out of those holes great abundance of fishes. And in other frozen seas are many suchlike, having black claws, living for the most part upon the seas, unless tempestuous weather drives them to the land.

Bears are bred in many countries, as in the Helvetian Alpine region, where they are so strong and full of courage that they can tear in pieces both oxen and horses. Likewise, there are bears in Persia, which do raven beyond all measure. So also the bears of Numidia, which are of a more elegant form and composition than the rest.

In the country of Arabia, from the promontory Dira to the South, are bears which live upon eating flesh, being of a yellowish color, and they far excel all other bears both in activity or swiftness and in quantity of body. Aristotle reports that there are white bears in Mysia, which, being eagerly hunted, do send forth such a breath that it putrefies immediately the flesh of the dogs, and whatsoever other beast comes within the savor thereof it makes the beast's flesh not fit to be eaten.

Thracia breeds white bears. The King of Ethiopia in his Hebrew Epistle which he wrote to the Bishop of Rome affirms that there are bears in his country. In Muscovia are bears of a snow-white, yellow, or dusky color.

A bear is of a most venerous and lustful disposition. Night and day, the females, with most ardent, inflamed desires, do provoke the males to copulation, and for this cause at that time they are most fierce and angry.

It is reported that, in the mountains of Savoy, a bear carried a young maid into his den by violence, where, in venerous manner, he had the carnal use of her body; and while he kept her in his den, he daily went forth and brought her home the best apples and other fruits he could get, presenting them unto her for her food in very amorous sort. But when he went to forage, he always rolled a huge, great stone upon the mouth of his den so that she could not escape. At length, her parents, after a long search, found her in the bear's den and delivered her from that savage and bestial captivity.

The time of the copulation of bears is usually in the beginning of winter, although it is sometimes in summer. The manner of their copulation is like a man's, the male moving himself upon the belly of the female, which lies on the earth flat upon her back, and they embrace each other with their forefeet. They remain a very long time in the act, inasmuch as if they were very fat at their first entrance, they disjoin not themselves again till they are made lean.

Immediately after they have conceived, they betake themselves to their dens, where without food they grow very fat (especially the males) only by sucking their forefeet. The males give great honor to the females great with young. Although the male and the female lie together in one cave, yet do they part it by a division or small ditch in the midst, neither of them touching the other. The nature of all bears is to avoid cold and therefore in the wintertime do they hide themselves, choosing rather to suffer famine than cold, lying for the most part three or four months together and never seeing the light.

When they first enter into their den, they betake themselves to quiet and rest, sleeping without any awakening for the first fourteen days. But how long the females go with young is not certain. Some affirm three months; others, but thirty days, which is more probable. At birth, the young ones are no bigger than rats, no longer than one's finger. It has been written that the young ones are littered without all form and fashion and are nothing but a little congealed blood like a lump of flesh. It has also been written that the old one frames the young ones to her own likeness with her tongue. These beliefs are false. Yet it is true that the young are littered blind and without hair, that their hinder legs are not perfect, that the forefeet are folded up like a fist, and that other members are deformed by reason of the immoderate humor or moistness in them. The female brings forth sometimes two and never above five. She daily keeps the young ones close to her breast, so warming them with the heat of her body and the breath of her mouth till they are thirty days old, at what time they come abroad, being in the beginning of May.

Some persons believe that bears eat the herb arum and that it causes them to pass away the whole winter in sleep. Concerning this herb, there is a pleasant but fabulous tale which is commonly believed.

According to the tale, there was a certain cowherd in the mountains of Helvetia, who, coming down a hill with a great caldron on his back, saw a bear eating of a root which he had pulled up with his feet. The cowherd stood still till the bear was gone and afterward came to the place where the beast had eaten and, finding more of the same root, did likewise eat it; he had no sooner tasted it but he had such a desire to sleep that he could not contain himself, but he had to lie down in the way and there fell asleep, having covered his head with the caldron to keep himself from the vehemency of the cold, and there slept all the wintertime without harm and never rose again till the springtime.

A bear is much subject to blindness, and for this cause they desire the hives of bees not only for the honey but because their eyes are cured by the stinging of the bees.

Bears are hunted and taken in divers ways. Some bears are killed in the mountains by poison, the country being so steep and rocky that hunters cannot follow them; some are taken in ditches of the earth and other traps. The inhabitants of Helvetia hunt bears with mastiff dogs and likewise shoot them with guns, giving a good sum of money to them that can bring them a slain bear. The Sarmatians plant a great many of sharp, pointed stakes under the trees where bees are bred, putting one stake hard into the hole where the bees go in and out. When a bear climbs the tree and comes to pull forth the stake in the hole, she becomes angry that the stake sticks so fast and with violence plucks it forth with both forefeet, whereby she loses her hold and falls down on the pointed stakes. Whereupon she dies, if she is not taken off.

There is a story of a man who, going to seek honey, fell into a hollow tree up to the breast in honey, where he lay two days, being not heard by any man; at length came a great bear to this honey, and when the bear put his head into the tree, the poor man took hold thereof, whereat the bear, suddenly affrighted, drew the man out of that deadly danger and so ran away for fear of a worse creature.

Bears are not easily tamed and not to be trusted, though they seem ever so tame. Yet bears are tamed for labors, and especially for sports, among the Roxolani and Lybians, being taught to draw water with wheels out of the deepest wells and to draw stones upon sleds to the building of walls.

A Prince of Lithuania nourished a bear very tenderly, feeding her from his table with his own hand, for he had accustomed her to be familiar in his court and to come into his own chamber when he wished. She would go abroad into the fields and woods, returning home again of her own accord; and she would, with her hand or foot, rub the Prince's chamber door to have it opened when she was hungry. It happened that certain young noblemen conspired the death of this Prince and came to the locked chamber door, rubbing it after the custom of the bear. The Prince, not fearing any evil and supposing it was his bear, opened the door, and the noblemen slew him immediately.

Bears care not for anything that is dead, and therefore, if a man can hold his breath as if he were dead, they will not harm him. This gave occasion to Aesop to write a fable of two companions and sworn friends, who, traveling together, met a bear. One of them ran away and got up into a tree; the other fell down and counterfeited himself dead, unto whom the bear came and smelt at his nostrils and ears for breath, but perceiving none, departed without hurting him. Soon afterward, the other friend came down from the tree and merrily asked his companion what the bear said in his ear.

"Marry," said he, "she warned me that I should never trust a fugitive friend as thou art, which didst forsake me in my greatest necessity."

Bears will bury dead bears, and it is received in many nations that children have been nursed by bears.

Heliogabalus was wont to shut up his drunken friends together, and suddenly in the nighttime he would put in among them bears, wolves, lions, and leopards, muzzled and disarmed, so that when his friends did awake, they should find such chamber-fellows as they could not behold without singular terror (if darkness did not blind them). And for this cause, many of them fell into swoons, sickness, ecstasy, and madness.

Vitoldus, King of Lithuania, kept certain bears to whom he cast all persons which spoke against his tyranny, putting them first of all into bearskins; his cruelty was so great that if he had commanded any of them to hang themselves, they would rather obey him than endure the terror of his indignation.

Pliny reports that, if a woman is in sore travail of childbirth, let a stone or an arrow which has killed a man, a bear, or a boar be thrown over the house wherein the woman is, and she shall be eased of her pain. If the blood or grease of a bear is set under a bed, it will draw unto it all the fleas and so kill them by cleaving thereunto. A bear's right eye dried to a powder and hung about children's necks in a little bag drives away the terror of dreams, and both the eyes whole, bound to a man's left arm, ease a quartan ague.

To conclude, the livers of a sow, a lamb, and a bear put together and trod to powder under one's shoes ease and defend cripples from inflammation. The stones in a perfume are good against the falling evil and the palsy, and that women may go their full time, they make amulets of bear's nails and wear them all the time they are with child.

The Bactrian Camel

THE CAMEL

TOPSELL DESCRIBES THE *two species of camels: the Bactrian or two-humped camel and the Arabian or one-humped camel. The physical description of the camel is accurate in many respects, but the lore concerning it is more fabulous than real. Although Topsell characterizes the camel as a hot-natured animal that is full of lust, he praises it for being modest about copulation and for abhorring "incest."*

There are divers kinds of camels, according to the countries wherein they are bred, as in India, in Arabia, and in Bactria.

All those which are in India are said to be bred in the mountains of the Bactrians and have two bunches on their back and one other on the breast, whereupon they lean. These Bactrian camels have sometimes a boar for a sire, which feeds with the flock of she-camels; for, as mules and horses will couple together in copulation, so also will boars and camels. And that a camel is so engendered sometimes, the roughness of his hair like a boar's or a swine's and the strength of his body are sufficient evidence.

A camel is called of the Grecians *dromos* by reason of his swiftness; and also an Arabian camel. The Arabian camel is of less stature but much swifter than the Bactrian, and whereas the Bactrian has two bunches on the back, the Arabian has but one. The Arabian camels are said to live fifty years. They were used for drawing chariots and for great presents for princes; and when they go to war, each one carries two archers, which sit upon him, back to back, shooting forth their darts, one against the front of the enemy and the other against the followers. Dromedaries are able to go a hundred miles in a day, bearing a burden of fifteen hundred weight; yea, sometimes two thousand. They bend upon the knee to take up load and rider, which received, they rise up again with great patience, being obedient and ruleable. Yet they kick when angry, but this is very seldom.

Camels have long and nimble necks, and in their neck toward the nether part of the throat, there is a place called *anhar* wherein a camel does by spear or sword most easily receive a mortal or deadly wound. A camel's belly is variable, now great, now small like an ox's. The tail is like the tail of an ass, hanging down to the knees. The foot is fleshy like a bear's, and therefore camels are shod with leather when they travel, lest the galling of their feet cause them to tire. Their manner of going or pace is like a lion's, so walking as the left foot never outgoes the right. All other beasts change the setting forward of their feet and lean upon their left feet while they remove their right, but camels alter step after step, so as the left foot behind, follows the right before, and the hind foot follows the left before.

Camels love grass and especially barley, which they eat up wonderfully greedily until all be in their stomach, and then will they chew thereupon all the night long, for they can ruminate and chew it so often as they please. They will not drink of clear or clean water but of muddy and slimy, and therefore they stamp in it with their feet. They will endure thirst for three or four days together, but when they come to drink, they suck in above measure, recompensing their former thirst and providing against that which is to come. Of all kinds of camels, the Bactrians are least troubled with thirst.

Camels are very hot by nature and therefore wanton and full of sport and wrath, braying most fearfully when they are angered. They engender like elephants and tigers: that is, the female lying or sitting on the ground, which the male embraces like other males. They continue in copulation a whole day together. When they are to engender, they go unto the most secret places they can find and make the procreation of their kind to be a most secret, honest action.

Unto this modesty for copulation we may add another divine instinct: the male will never copulate with his mother or his sister. It is sincerely reported that a certain camel-keeper (desirous to try this secret) covered a female camel in all parts of her body except her privy parts, so that nothing could be seen of her, and then he brought her son to mate with her, which in the rage of lust he performed. As soon as he had done this, the master and owner pulled away the disguise from the mother in the presence of the son, and he instantly perceived his keeper's fraud in making him unnaturally to have copulation with his own mother. In revenge, he ran upon the keeper, and, taking him in his mouth, lifted him up into the air and immediately let him fall with noise and cry underneath his murdering and man-quelling feet, where, with unappeasable wrath, he pressed and trod to pieces the incest marriage-causer between him and his dearest mother. Yet not satisfied with this, like some reasonable creature deprived of heavenly grace and carried with deadly revenge against such uncleanness, being persuaded that the guilt of such an offense could never receive sufficient expiation by the death of the first deviser, he adjudged himself to death; and, running to and fro, at last he found a steepy rock, from whence he leaped down to end his life. Although he could not prevent his offense against his mother, he thought it best to cleanse away his mother's adultery with the sacrifice of that blood which was first conceived in that womb which he had defiled.

Camels are used for carriage, which they will perform with great facility. Their keepers teach them to kneel and lie down to take up their burdens, which, by reason of their height, a man cannot lay on them. One Bactrian camel has carried above ten pounds of corn and above that a bed with five men therein. Camels will travel in a day about forty ordinary miles.

They are used for plow in Numidia, and for this cause they are yoked sometimes with horses. Heliogabalus, like the Tartarians, yoked them together not only for private spectacles and plays but also for drawing of wagons and chariots.

It is forbidden in Holy Scripture to eat a camel, for although it chews the cud, the hoof is not altogether cloven. And, besides, the flesh is hard of digestion, and the juice very naught, heating the body above measure. Yet many times have men of base condition and minds eaten thereof. It is reported that the King of Persia was wont to have a whole camel roasted for his own table at his royal feastings, and Heliogabalus likewise caused to be prepared for himself the heels of camels and the spurs of cocks and hens, saying (though falsely) that God commanded the Jews to eat them.

The camel is a disdainful and discontented creature. For this cause, it is feigned of the poets that camels sought Jupiter to give them horns, with which petition he was so offended that he took from them their ears.

In the lake of Asphaltites, wherein all things sink that come in it, many camels and bulls swim through without danger. The Arabians sacrifice a camel to the unknown God, because camels go into strange countries, and likewise they sacrifice their virgins before they are married because of the chastity of this beast.

Camels are hated of horses and lions. When Xerxes traveled over the river Chidorus, through Paeonia and Crestonia, in the nighttime the lions descended into the camp and touched no creatures except the camels, whom they destroyed for the most part.

A camel will live fifty or a hundred years in the soil wherein he is bred; and, if he is translated into any other nation, he falls into madness or scabs or the gout, and then he lives not above thirty years.

There are medicinal properties in camels. If a man infected with poison be put into the warm belly of a camel newly slain, this loosens the power of the poison and gives strength to the natural parts of the body. The fat, taken out of the bunch and perfumed, cures the hemorrhoids; and the brain dried and drunk with vinegar helps the falling evil; the gall drunk with honey helps the quinsy; and, if it is laid to the eyebrows and forehead, boiled in three cups of the best honey, it cures the dimness of the eyes and avoids the flesh that grows in them. If the hairs of a camel's tail be wound together like a string and tied to the left arm, they will deliver one from a quartan ague.

The urine is most profitable for running sores. There have been those who have preserved it five years and used it against hardness of the belly; washing also therewith sore heads; and it helps one to the sense of smelling if it is held to the nose; likewise against the dropsy, the spleen, and the ringworm.

The Two Sorts of Camelopardals

THE CAMELOPARDAL

THE CAMELOPARDAL IS *the giraffe. Topsell's description contains some inaccuracies, but in general it gives a fairly good idea of what a giraffe looks like. The two illustrations are from Gesner's* Historia. *Although the caption accompanying the illustrations says that they show the two kinds of camelopardals, Topsell does not describe two kinds.*

Some writers thought that this animal was the offspring of a male camel and a female panther (the view expressed by Topsell); others did not regard the camelopardal as a hybrid offspring but as a separate kind of animal. At any rate, the camelopardal was usually said to have markings like a panther's, a head like a camel's, and other parts like those of two or three other animals.

This beast is called in Hebrew *zamer,* which the Arabians translate *saraphah* and sometimes *gyrapha, gyraffa,* and *zirafa;* the Chaldeans, *deba* and *ana;* the Persians, *seraphah;* and the Septuagint Grecians, *camelopardalis,* which word is also retained by the Latins.

There were ten of these beasts seen at Rome in the days of the Emperor Gordianus, and before that time when Caesar was dictator. And such a beast was sent by the Sultan of Babylon to the Emperor Frederick.

This beast is engendered of a camel and a female libbard. (The same beast which the Latins call *panthera,* the Grecians call *pardalis.*) The head of the camelopardal is like a camel's, his neck like a horse's, and his body like a hart's; and his cloven hooves are the same as a camel's. The color of this beast is for the most part red and white mixed together, therefore very beautiful to behold because of the variable and interchangeable skin being full of spots. But they are not always of one color.

The camelopardal has two little horns of the color of iron growing on his head. His mouth is small like a hart's. His tongue is nearly three feet long, and with that he will so speedily gather in his food that the eyes of man will fail to behold his haste. His neck is diversely colored and is fifteen feet long. He holds up his neck higher than a camel's and far above the proportion of his other parts. His forefeet are much longer than his hind feet, and therefore his back declines toward the buttocks, which are very much like those of an ass. The pace of this beast differs from that of all other beasts in the world, for he does not move his right and left foot one after another but both together, and so likewise the other, whereby his whole body is removed at every step.

These beasts are plentiful in Ethiopia, India, and the Georgian region which was once called Media. It is a solitary beast and keeps altogether in woods, if it not be taken when it is young. They are very tractable and easy to be handled, so that a child may lead them with a small line or cord about their head, and, when any come to see them, they willingly and of their own accord turn themselves round as it were of purpose to show their soft hairs and beautiful color, being, as it were, proud to ravish the eyes of the beholders.

The skin is of great price and estimation among merchants and princes, and it is said that underneath the belly the colored spots are wrought in fashion of a fisher's net, and the whole body so admirably intercolored with variety that it is in vain for the wit or art of man to go about to endeavor the emulous imitation thereof. The tail of the beast is like the tail of an ass, and I cannot judge that the camelopardal is either swift for pace or strong for labor. The flesh is good for meat and was allowed to the Jews by God for a clean beast.

The Cat

THE CAT

THE HISTORY OF *the cat is one of Topsell's most interesting histories. Topsell depicts the cat as a creature of many moods and shows its playfulness, independence, and mysteriousness. In a famous observation about his cat, French essayist Montaigne wondered whether it was not amusing itself with him more than he was with it. Topsell seems to be of the opinion that the cat is the one that finds the greater sport and amusement.*

As Topsell shows, the cat was venerated in ancient times. In ancient Egypt, cats were worshiped, and people were put to death for killing them. In Topsell's time, people valued the cat for killing mice and rats, but they also regarded the cat with suspicion, because they associated it with witchcraft and they believed that it was a carrier of diseases. Thus, Topsell calls the cat "a dangerous beast" and cautions people to make "more account of its use than of its person."

Ovid says that, when the Giants warred with the gods, the gods put upon them the shapes of beasts, and the sister of Apollo lay for a spy in the likeness of a cat, for a cat is a watchful and wary beast seldom overtaken and most attendant to her sport and prey. The Egyptians considered cats hallowed beasts and kept them in their temples. It is known also that it was capital among them to kill an ibis, an asp, a crocodile, a dog, or a cat. The Arabians worshiped a cat for a god, and when the cat died, they mourned as much for her as for the father of the family, shaving the hair from their eyelids and carrying the beast to the temple, where the priests salted it and gave it a holy funeral in bubastum (which was a burying place for cats near the altar).

Once cats were all wild, but afterward they retired to houses, wherefore there are plenty of them in all countries.

A cat is in all parts like a lioness except in her sharp ears. Her flesh is soft and smooth: her eyes glister above measure, especially when a man comes to see them on the sudden, and in the night they can hardly be endured for their flaming aspect. Albertus compares their eyes to carbuncles in dark places because in the night they can see perfectly to kill rats and mice. The root of the herb valerian (commonly called *phu*) is very like to the eye of a cat, and wheresoever it grows, if cats come thereunto, they instantly dig it up for the love thereof, as I myself have seen in mine own garden, and not once only but often, even when I had caused it to be hedged or compassed round about with thorns. It smells marvelously like a cat.

The Egyptians have observed in the eyes of a cat the increase of the moonlight, for with the moon they skin more fully at the full and more dimly in the change and wane, and the male cat does also vary his eyes with the sun, for when the sun arises, the apple of his eye is long; toward noon it is round, and at the evening it cannot be seen at all, but the whole eye shows alike.

The teeth of a cat are like a saw, and if the long hairs growing about her mouth are cut away, she loses her courage. Her nails are sheathed like the nails of a lion. She strikes with her forefeet both dogs and other things, as a man does with his hand.

This beast is wonderful nimble, setting upon her prey like a lion by leaping, and therefore she hunts both rats and all kinds of mice and birds, eating not only them but also fish, wherewithal she is best pleased. Having taken a mouse, she first plays with it and then devours it, but her watchful eye is most strange, to see with what pace and soft steps she takes birds and flies.

Some have said that a cat will fight with serpents and toads and kill them, and if she perceives that she is hurt by them, she immediately drinks water and is cured. But I cannot consent unto this opinion.

Pontzettus shows that cats and serpents love one another, for there was (says he) in a certain monastery a cat nourished by the monks, and suddenly most of the monks which were used to play with the cat fell sick. The physicians could find no cause but some secret poison, and all of the monks were sure they had never tasted any. At the last, a poor laboring man came unto them, affirming that he saw the abbey cat playing with a serpent. The physicians conceived that the serpent had emptied some of his poison upon the cat, and she had brought it to the monks, who by stroking and handling her had been infected therewith. There remained one difficulty: namely, how it came to pass the cat herself was not poisoned. It was resolved that, as the serpent's poison came from him but in play and sport, and not in malice and wrath, the venom being lost in play, the poison neither harmed the cat at all nor much endangered the monks.

Cats will hunt apes and follow them to the woods. In Egypt, certain cats set upon an ape, who immediately took himself to his heels and climbed up into a tree, and the cats followed him with the same celerity and agility (for they can fasten their claws into the bark and run up very speedily). Seeing himself overmatched with the number of his adversaries, he leaped from branch to branch and at last took hold of the top of a bough, where he did hang so ingeniously that the cats did not dare approach unto him for fear of falling and so departed.

The nature of this beast is to love the place of her breeding; neither will she tarry in any strange place, although carried far. She is never willing to forsake the house for the love of any man, and this is contrary to the nature of a dog, who will travel abroad with his master. Although their masters forsake their houses, yet will not cats bear them company, and being carried forth in close baskets or sacks, they will return again or lose themselves.

A cat is much delighted to play with her image in a glass, and if at any time she beholds her image in water, she immediately leaps down into the water, which she naturally does abhor; but if she is not quickly pulled forth and dried, she dies because she is impatient of all wet.

Nothing is more contrary to the nature of a cat than is wet and water, and for this cause came the proverb that they love not to wet their feet.

The cat is a neat and cleanly creature, oftentimes licking her own body to keep it smooth and fair (having naturally a flexible back for this purpose) and washing her face with her forefeet. But some people observe that if she puts her feet beyond the crown of her head it is a presage of rain, and some say that if the back of a cat is thin the beast is of no courage or value.

Cats love fire and warm places, whereby it often falls out that they often burn their coats. They desire to lie soft, and in the time of their lust (commonly called caterwauling) they are wild and fierce, especially the males, who at that time (unless they be gelded) will not keep the house. At that time they have a special, direful voice. The females are above measure desirous of procreation, for which cause they provoke the male, and if he does not yield to their lust, they beat and claw him, but it is only for love of young and not for lust. The male is more libidinous, and seeing that the female will never more engender with him during the time her young ones suck, he kills and eats them if he meets with them, for when she is deprived of her young, she seeks out the male of her own accord. For this cause, the female most warily keeps her young ones from his sight.

When they have littered, or as we commonly say kittened, they rage against dogs and will suffer none to come near their young ones. The best of the young ones to keep are such as are littered in March.

Cats cannot abide the savor of ointments but fall mad thereby. They are sometimes infected with the falling evil but are cured with gobium.

It is needless to spend any time about the cat's loving nature to man: how she flatters by rubbing herself against one's legs and how she has as many tunes as turns in her voice, for she has one voice to beg and to complain, another to testify her delight and pleasure, and another among her own kind, flattering, hissing, puffing, and spitting, insomuch that some people have thought that cats have a peculiar, intelligible language among themselves. How she begs, plays, leaps, looks, catches, tosses with her foot, sometimes creeping, sometimes lying on the back, playing with one foot and sometimes on the belly, snatching now with mouth and anon with foot.

As this beast has been familiarly nourished of many, so have they paid dear for their love, being requited with the loss of their health and sometimes of their life for their friendship, and this happens worthily, for those who love any beast in a high measure have so much the less charity unto man.

It must be considered what harms and perils come unto men by this beast. It is most certain that the breath and savor of cats consume the radical humor and destroy the lungs, and they who keep their cats with them in their beds have the air corrupted and fall into hectics and consumptions. There was a certain company of monks much given to nourish and play with cats, whereby they were so infected that, within a short space, none of them were able to say, read, pray, or sing in all the monastery.

Cats are also dangerous in the time of pestilence, for they are not only apt to bring home venomous infection but to poison a man with the very looking upon him. There is in some men a natural dislike and abhorring of cats, the natures of these men being so composed that not only when they see cats but the cats being near them and unseen and hidden of purpose, they fall into passions, fretting, sweating, pulling off their hats, and trembling fearfully.

The familiars of witches do most ordinarily appear in the shape of cats, which is an argument that this beast is dangerous to soul and body.

The hair of a cat being eaten unawares stops the artery and causes suffocation; and I have heard that, when a child has gotten the hair of a cat into his mouth, it has so cloven and stuck to the place that it could not be gotten off again, and has in that place bred either the wens or the king's evil. To conclude this point, it appears that this is a dangerous beast and that therefore as for necessity we are constrained to nourish it for the suppressing of small vermin, so with a wary and discreet eye we must avoid its harms, making more account of its use than of its person.

The Cockatrice

THE COCKATRICE

IN ANCIENT TIMES, *this creature was identified as the basilisk serpent. The basilisk was thought to be the king of serpents, generated in the same way as were other land serpents. The belief that the creature was derived from the "egg" of an old cock originated in the Middle Ages. (A concretion that has the appearance of an egg is sometimes formed in old roosters.) A confusion about etymology arose, and the creature came to be called a cockatrice as well as a basilisk. This animal was customarily depicted as having the head, wings, and feet of a rooster and the tail of a serpent.*

Topsell did not believe that the basilisk, or cockatrice, was a creature compounded of a rooster and a serpent. He thought that it was a serpent and that its manner of generation was like that of other land serpents. He believed that a worm resembling the basilisk might be hatched from a cock's "egg," but such a worm was not the creature he was describing in his history.

(For an excellent discussion of the basilisk, see T. H. White, The Bestiary: A Book of Beasts *[New York: G. P. Putnam's Sons, 1954]).*

This beast is called by the Grecians *basiliscos* and by the Latins *regulus* because he seems to be the king of serpents. There are many other serpents bigger than he is, and therefore it is not because of his magnitude or greatness that he has been looked upon as the king of serpents. Instead, it is because of his stately pace and magnanimous mind, for he does not creep on the earth like other serpents but goes half upright, for which reason all other serpents avoid his sight. And it seems nature has ordained him as a king; for besides the strength of his poison, which is incurable, he has a certain comb or coronet upon his head.

There is some question among writers about the generation of this serpent. Some writers (and those very many and learned) affirm him to be brought forth of a cock's egg. For they say that, when a cock grows old, he lays a certain egg without a shell. Instead, it is covered with a very thick skin which is able to withstand the greatest force of an easy blow or fall. They also say that this egg is laid only in the summertime, about the beginning of dog days, and that the egg is not as long as a hen's egg but is round and orbicular. They say that the egg is generated of the putrefied seed of the cock and that afterward is sat upon by a snake or a toad, and that the cockatrice is thus brought forth, being half a foot in length, with the hinder part like a snake and the front part like a cock. The common opinion of Europe is that the egg is nourished by a toad and not by a snake.

I am persuaded that, when a cock grows old and ceases to tread his female in the ordinary course of nature, which is in the seventh or ninth year of his age, or at the most in the fourteenth, there is a certain concretion bred within him by the putrefied heat of his body, through the staying of his seed generative. This concretion hardens into an egg and is covered with such a shell as is said already. Being nourished by the cock or some other beast, this egg brings forth a venomous worm such as bred in the bodies of men, or as wasps, horseflies, and caterpillars are engendered of horse dung or other putrefied humors of the earth. Therefore, out of this egg may a venomous worm proceed, and in proportion of body and pestiferous breath it may resemble the African cockatrice or basilisk and may be one kind of cockatrice, yet it is not the same beast whereof we purpose here to treat, for that beast is not hatched of a cock's egg but is generated like other serpents of the earth.

The Holy Scriptures make mention of the cockatrice and her eggs, and there are many grave writers who affirm that there are cockatrices and that they infect the air and kill with their sight.

The length of this serpent must be three or four feet at the shortest. Else how would it be such a terror to other serpents or how could the forepart of it arise so eminently above the earth if the head were not lifted at the least a foot from the ground? So then we will take it for granted that this serpent is as big as a man's wrist and the length of it answerable to that proportion.

Because of his conceived generation from a cock, many have described the cockatrice in the forepart to have wings and in the hinder part to have a tail like a serpent. But we do not find sufficient authority to warrant such a belief.

The eyes of the cockatrice are red or somewhat inclining to blackness. The skin and carcass have been accounted precious. We read that the inhabitants of the city of Pergamum bought certain pieces of a cockatrice and gave for it two pounds and a half of silver. And because there is the opinion that no bird, spider, or venomous beast will endure the sight of this serpent, they did hang up the skin stuffed, in the temples of Apollo and Diana, in a certain thin net made of gold. Therefore it is said that never any swallow, spider, or any serpent would come within those temples. | Not only the skin or the sight of the cockatrice works this effect, but also its flesh being rubbed upon the pavement, posts, or walls of any house. And moreover, if silver be rubbed over with the powder of the cockatrice's flesh, it is said that it gives it a tincture like unto gold.

The hissing of the cockatrice, which is its natural voice, is terrible to other serpents, and as soon as they hear it, they prepare themselves to fly away. We read that many times in Africa mules fall down dead from thirst, or else lie dead on the ground for some other causes, and unto these carcasses innumerable troops of serpents gather themselves to feed. When the basilisk smells the said dead bodies, he gives forth his voice, and at the first hearing of it, all the other serpents hide themselves in the adjoining sands or else run into their holes, not daring to come forth until the cockatrice has well dined and satisfied himself, at which time he gives another signal by his voice of his departure: then come the other serpents forth but never dare meddle with the remnants of the dead beasts but go away to seek some other prey.

And if it happens that any other pestiferous beast comes unto the waters to drink near the place wherein the cockatrice is lodged, so soon as it perceives the cockatrice's presence (though the cockatrice be not heard nor seen), the beast departs back again without drinking, neglecting its own nutriment to save itself from further danger.

Although the hissing of the cockatrice is terrible to other serpents, and his breath and poison are mortal to all manner of beasts, yet has God not left this vile serpent without an enemy in nature. For the weasel and the cock are his triumphant victors. When people take weasels, they find the caves and lodging places of cockatrices by discerning where the upper face of the earth has been burned with the hot poison of the cockatrices. They put the weasel into the cockatrice, and at the sight whereof the cockatrice flies like a weakling overmatched with too strong an adversary, but the weasel follows after her and kills her. Yet this is to be noted: both before the fight and after the slaughter, the weasel arms herself by eating rue, or else she would be poisoned with the contagious air about the cockatrice.

The basilisk is not only afraid of the sight of a cock but becomes almost dead when he hears him crow. This is well-known throughout Africa, and therefore all travelers who go through the deserts take with them a cock for their safe conduct against the poison of the basilisk.

Now we are to treat of the poison of this serpent. It is a hot and venomous poison, infecting the air around so that no other creature can live near him, for he kills not only by his hissing and by his sight but also by his touching, both immediately and mediately. That is to say, not only when a man touches the body itself but also when he touches a weapon wherewith the body was slain or any other dead beast slain by it. There is a common story that a horseman taking up a spear which had been thrust through a cockatrice did not only draw the poison of the cockatrice into his own body and so died, but also killed his horse thereby.

The question is in what part of the body the poison lies. Some say in the head alone, but this seems not to be true, for the basilisk kills with the fume of the whole body, and, besides, when it is dead it kills anyone who touches it, and the man or beast so slain does also kill another by touching it. Some say that the poison is in the breast and that the basilisk breathes at the sides and at many other places of the body, through and between the scales. It is true that it does so breathe, for otherwise the burning fume that proceeds from this poisonful beast would burn up the entrails if it came out of the ordinary place. Therefore Almighty God has so ordained that it should have spiraments and breathing places in every part of its body to vent away the heat, lest in very short time by inclusion thereof the whole body should be utterly dissolved and one part separated from the other.

But, seeing that it is most manifest that the poison is universal in the body, we will leave inquiry about the seat of the poison and dispute of its instruments and effects.

The cockatrice kills his own kind by sight, hearing, and touching. By his own kind, I mean other serpents and not other cockatrices, for they can live one beside another. But there is a question whether the cockatrice dies by the sight of himself. Some have affirmed so much, but I dare not subscribe thereunto, because, in reason, it is impossible that anything should hurt itself if it does not hurt another of its own kind. Yet, if in the secret of nature God has ordained such a thing, I will not strive against them that can show it.

And therefore I cannot without laughing remember the old wives' tale of the cockatrices that have been in England, for I have oftentimes heard it related confidently that once our nation was full of cockatrices and that a certain man did destroy them by going up and down in a glass, whereby their own shapes were reflected upon their own faces and so they died. This fable is not worth refuting, for it is more likely that the man should first have died by the corruption of the air from the cockatrices than the cockatrices to die by the reflection of their own similitude from the glass, unless it could be shown that the poisoned air could not enter into the glass wherein the man did breathe.

The poison of the cockatrice infects the air, and the air so infected kills all living things and likewise all green things, fruits, and plants of the earth. The cockatrice burns up the grass upon which it goes or creeps, and the fowls of the air fall down dead when they come near its den or lodging. Sometimes the cockatrice bites a man or a beast, and by that wound the blood turns into choler, and so the whole body becomes yellow as gold, killing all that touch it or come near it.

Nile Crocodile

THE CROCODILE

TOPSELL DIVIDES CROCODILES *into two groups: water living and land living. The water crocodiles are the common crocodile and the crocodile of the Nile. Land crocodiles are the Arabian or Egyptian crocodile, the crocodile of Bresilia (Brazil), and the scink, or skink. He includes a description of each of these three kinds, but these descriptions have been omitted in this book.*

Many details about the appearance and habits of crocodiles are correct. Topsell has much to say about the worship of the crocodile among the ancient Egyptians; his reasons why the crocodile was considered divine appear to have a basis in fact. The history creates a vivid picture of the fierceness and rapacity of the crocodile.

There are many kinds of crocodiles, and people distinguish them into the crocodiles of the earth and the water. Those of the earth are subdivided into the Arabian or Egyptian land crocodile, the land crocodile of Bresilia, and the scinus or scink. The crocodiles of the water are subdivided into the common crocodile and that of the Nile.

In speaking of the crocodile, Aristotle for the most part calls it *aquatilis* and *fluviatilis*. Yet it is not to confine it to the waters and rivers as though it never came out of them, but it is only to note that particular kind of crocodile from the crocodile of the earth. The crocodiles of the water live on the land and in the water. In the daytime, they abide on the land because the earth is hotter than the water, but in the night, when the water is warmer than the earth, they live in the water. While they are on the land, they are delighted with the sunshine and lie therein so immovable that a man would take them to be stark dead.

In the next place, we are to consider the countries wherein crocodiles are bred and keep their habitation, and those are especially Egypt, for that only has crocodiles of both the water and the land. The crocodiles of the Nile are amphibia and live in both elements. Crocodiles are not only in the Nile but also in the pools near adjoining. The river Bambotus near to Atlas in Africa does also bring forth crocodiles, and Pliny reports that in Darat, a river of Mauritania, there are crocodiles engendered. Apollonius reports that, when he passed by the river Indus, he met with many sea horses and crocodiles, such as are found in the Nile, and besides these countries I do not remember any other wherein are engendered crocodiles of the water, which are the greatest and most famous of all crocodiles.

The crocodiles of the earth, which are of lesser note and quantity, are more plentiful, for they are found in Lybia, and in Bythinia, and in the mountain Syagrus in Arabia, and in the woods of India.

The crocodiles of the earth are not above two cubits long, or sometimes eight at the most, but the water crocodiles are sixteen and sometimes more.

The color of a crocodile is like to saffron, that is, between yellow and red, more inclining to yellow than red. Peter Martyr says that their belly is somewhat whiter than the other parts.

Now, a crocodile is like a lizard in all points except its size and its tail. The body of a crocodile is rough all over, being covered with a certain bark or rind so thick, firm, and strong as it will not yield (especially about the back) unto a cartwheel when the cart is loaded, and the upper parts and the tail are impenetrable to any dart or spear, and scarcely to a pistol or small gun. But the belly is softer, whereon he receives wounds with more ease.

The covering of the back is distinguished into divers divided shells, standing up far above the flesh, and towards the sides they are less eminent, but on the belly they are more smooth, white, and very penetrable. The head is very broad, and the snout is like a swine's. When he eats or bites, he never moves his nether or under chap.

The eyes of the crocodiles of the water are reported to be like unto a swine's. In the water, they see very dimly, but out of the water they are sharp-sighted. They have but one eyelid, and that grows from the nether part of the cheek. The Egyptians say that only the crocodile among all living creatures in the water draws a certain thin, bright skin from his forehead over his eyes, wherewithal he covers his sight; and this I take to be the only cause of his dim sight in the waters.

The crocodile has no tongue nor so much as any appearance of a tongue. It has great teeth standing out; all of them stand out before visibly when the mouth is shut, and fewer behind. These teeth are white, long, sharp, and a little crooked and hollow. The tail is of the same length that the whole body has, and it is rough and armed with hard skin upon the upper part and the sides but beneath is smooth and tender. It is doubtful whether the crocodile has any place of excrement except the mouth.

THE CROCODILE

The tail of a crocodile is his strongest part, and they never kill any beast or man but first of all they strike him down and astonish him with their tails, and for this cause the Egyptians by a crocodile's tail do signify death and darkness.

They devour both men and beasts if they find them in their way or near the banks of the Nile wherein they abide, taking sometimes a calf from the cow and carrying it whole into the waters. The dogs of Egypt by a kind of natural instinct do not drink but as they run, for fear of the crocodiles.

When they desire fishes, they put their heads out of the water as it were to sleep, and then suddenly when they espy a booty, they leap into the waters upon them and take them. When the Nile is lowest and sunk into the channel, then the crocodiles in the waters do grow most hungry because the fish are gone away; and then the subtle beast will cover himself over with mud or sand and so lie in the bank of the river, where he knows that women come to fetch water or the cattle to drink, and when he espies his advantage, he suddenly takes the woman by the hand with which she takes up water and draws her into the river, where he tears her in pieces and eats her. In like sort, he deals with oxen, cows, asses, and other beasts. If hunger forces him to the land and he meets with a camel, horse, ass, or suchlike beast, then with the force and blows of his tail he breaks his legs, and so laying him flat on the ground, kills and eats him, for so great is the strength of a crocodile's tail that it has been seen that one stroke of it has broken all four legs of a beast at one blow.

Crocodiles are exceedingly fruitful. They bring forth every year and lay their eggs in the earth or dry land. During the space of threescore days, they lay every day an egg; within the like space, the young ones are hatched by sitting or lying upon the eggs by course, the male for a while and the female for a while. The egg is not much bigger than the egg of a goose, and the young one out of the shell is of the same proportion. And so from such a small beginning does this huge and monstrous serpent grow to his great stature.

When the crocodile has laid her eggs, she carries them to the place where they shall be hatched, for, by a natural providence and foresight, she avoids the waters of the Nile and places her eggs beyond the compass of the floods. By observation of this, the people of Egypt know every year the inundation of the Nile before it happens. In the measure of this place, it is apparent that this beast is not indued only with a spirit of reason but also with a prophetical geographical delineation, for so she places her eggs in the brim or bank of the flood (before the flood comes) that the water may cover the nest but not herself that sits on the eggs.

The crocodile abhors all manner of noise, especially from the strained voice of a man, and where a crocodile finds himself valiantly assaulted, there also is he discouraged.

The inhabitants of the island Tentyrus within the Nile are the greatest terror unto crocodiles, for these people make the crocodiles run away with their voices and many times pursue and take them in snares. According to Solinus, the Tentyrites are of a most adverse nature to the crocodile; and although their persons or presence be of small stature, yet herein is their courage admired because at the sudden sight of a crocodile they are no whit daunted: for one of them will dare to meet the beast and provoke it to run away. They will also leap into the rivers and swim after the crocodile, and meeting with it, they cast themselves upon the beast's back without fear, riding on it as upon a horse. And if the beast lifts up its head to bite, the man puts into its mouth a wedge when it gapes, holding the wedge hard at both ends with both hands, and so as it were with a bridle, the man will lead, or rather drive, the crocodile a captive to land, where the people with their noise will so terrify it that they make it cast up any bodies which it had swallowed into its belly.

Strabo has recorded that, when crocodiles were brought to Rome, these Tentyrites followed and drove them. There was a certain great pool or fishpond walled about, except one passage for the beast to come out of the water into the sunshine: and when the people came to watch, the Tentyrites would draw the crocodiles to the land with nets and would put them back again into the water at their own pleasure. For the Tentyrites so hook the crocodiles by their eyes and the bottom of their bellies (which are their tenderest parts) that, like horses broken by their riders, the crocodiles yield unto them and forget their strength in the presence of these their conquerors.

In divers places in Egypt, crocodiles were worshiped by the people. The reasons of divine worship or honor given to the crocodile are worth the noting so that the diligent reader may the better have some taste of that ancient blindness by which our forefathers were misled and seduced to forsake the most glorious and ever-blessed principles of divinity for arguments of no weight.

First, the idolatrous priests thought that there was some divine power in the crocodile because it has no tongue. The Deity has no need of a voice to express His meaning and speaks in action though not in voice, even as all that is in the crocodile is action and not voice.

Secondly, the crocodile was accounted divine by reason of the certain thin, smooth skin coming from the midst of his forehead and covering his eyes, so that, when he is thought to be blind, yet he sees. Even so is it with divine power, for even when it is not seen, yet it does see perfectly all mortal things.

Again, crocodiles were honored with divine power because they usually foreshow the overflowing of the Nile by their eggs and nests. Thereby, the husbandmen know when to till their land and when not to sow and plant, and when to lead forth their flocks and when not to. This benefit is also ascribed to divinity.

Again, the crocodile lays threescore eggs and lives threescore years, which number of threescore was, in ancient times, the first dimension of heaven and heavenly things.

There is a tale in Diodorus Siculus of the origin of the divine worship of a crocodile, and although this tale cannot be but fabulous, yet I have thought good to insert it in this place to show the vanity of superstition and idolatry. There was a king of Egypt called Minas, or as Herodotus calls him Menes, who, following his hounds in hunting into a certain marsh, fell in with his horse and there stuck so fast that none of his followers dared to come after him to release him. He would have perished had not a crocodile come and taken him up upon his back and set him safe upon the dry land. Because of this miracle, the said king there built a city and caused a crocodile to be worshiped. This crocodile was called Sychus by all the inhabitants of the city, and the king also gave all of the marsh for the sustenance of the crocodile. It was nourished with bread, flesh, wine, cakes, boiled flesh, and sweet new wine: so that when any man came to the lake where it was kept, the priests would call the beast out of the water, and when it had come to the land, one of them opened its mouth, and the other put in meat, delicacies, and wine.

Yet even the Egyptians themselves account the crocodile a savage and murdering beast. The common proverb about the crocodile's tears justifies the treacherous nature of this beast, for there are not many brute beasts that can weep, but such is the nature of the crocodile that, to get a man within his danger, he will sob, sigh, and weep, as though he were in extremity, but suddenly he destroys him. Others say that the crocodile weeps after he has devoured a man. Howsoever it be, the proverb notes the wretched nature of hypocritical hearts which beforehand will with feigned tears endeavor to do mischief or else after they have done it will be outwardly sorry, as Judas was for the betraying of Christ before he went and hanged himself.

The males do love their females above measure. Yea, even to jealousy. Once there were certain mariners who saw two crocodiles together in carnal copulation upon the sands near the river. The greedy mariners forsook their ship and betook themselves to a long boat, and with great shouting, hollering, and crying, made towards them in very courageous manner: the male at the first assault fell amazed and, greatly terrified, ran away as fast as he could into the water, leaving his female lying upon her back (for when they engender, the male turns her upon her back because she cannot do it herself on account of the shortness of her legs). So the mariners finding her upon her back and not able to turn herself over slew her easily and took her away with them.

Soon after, the male returned to the place to seek his female, but not finding her and perceiving blood upon the sand conjectured truly that she was slain. He immediately cast himself into the Nile again, and in his rage swam stoutly against the stream until he overtook the ship wherein his dead female was. He immediately set upon it, lifting himself up and catching hold of the sides, and he would certainly have entered it had not the mariners with all their force battered his head and hands with clubs and staves until he was wearied and forced to give over his enterprise, and so with great sighing and sobbing, he departed from them. From this relation, it is most clear what natural affection they bear each other and how they choose their fellows, as it were fit wives and husbands for procreation.

And it is no wonder that they make much of one another, for besides themselves they have no friends except swine and a small bird called the *trochilus*.

The trochilus follows crocodiles for the benefit of its own belly. When the crocodile eats greedily, there sticks fast in his teeth some part of his prey, which troubles him very much and many times engenders worms. Then the beast to help himself comes on land and lies gaping against the sunbeams westward. The bird perceiving it flies to the jaws of the beast and there first with a kind of tickling-scratching procures (as it were) license of the crocodile to pull forth the worms, and so eats them all out and cleanses the teeth thoroughly, for which cause the beast is content to permit the bird go into his mouth.

It is said by some that when all is cleansed, the ungrateful crocodile endeavors suddenly to shut his chaps together upon the bird and to devour his friend, like a cursed wretch that makes no reckoning of friendship, but the turn served, requites good with evil. But nature has armed this little bird with sharp thorns upon her head, so that when the crocodile endeavors to shut his chaps and close his mouth upon it, those thorns prick him in his palate so that full sore against his unkind nature he lets the bird fly safe away.

Yet there are many kinds of trochili. Some do not have sharp thorns on their heads, and there being nothing to stop the closing of the crocodile's mouth, they are devoured. Therefore this enforced amity between the bird and the crocodile is to be understood of only the *cledororynchus,* which has thorns on the head.

Some writers affirm that the crocodile destroys all kinds of these birds without exception. Others say that he destroys none, but that when he feels his mouth sufficiently cleansed, he wags his upper chap as it were to give warning of avoidance, and in favor of the good turn, he lets the bird fly away at his own pleasure.

It is more likely to be true that the crocodile endeavors to devour all of these birds. Leo Africanus says that it was the constant and confident report of all Africa that the crocodile devours all of them except the *cledororynchi,* which it cannot do because of the thorns upon the head.

That there is an amity and natural concord between swine and crocodiles is also gathered, because only swine among all other four-footed beasts do without danger dwell and feed upon the banks of the Nile, even in the midst of crocodiles; and therefore it is probable that they are friends in nature.

But, oh, how small a sum of friends has the crocodile, and how unworthy of love among all creatures that only two, in heaven or earth, air or water, will adventure to come near him; and one of these, which is the best deserving, the crocodile devours and destroys, if he gets it within his power.

Seeing that the friends of the crocodile are so few, the enemies of it must be many. In the first rank of enemies is the ichneumon or Pharaoh's mouse, who rages against their eggs and their persons, for it is certain that the ichneumon hunts with all sagacity of sense to find out their nests, and having found them, she spoils, scatters, breaks, and empties all their eggs. Strabo reports that the ichneumon also watches the old ones asleep, and if she finds one of them with his mouth open against the beams of the sun, she suddenly enters the mouth, and, being small, creeps down the crocodile's vast and large throat before the beast is aware. Then she puts the crocodile to exquisite and intolerable pain by eating the guts and the soft belly asunder. The crocodile tumbles to and fro sighing and weeping, now in the depths of the water, now on the land, never resting till his strength fails. The enemy within sports herself in the consumption of the vital parts, which waste and wear away by yielding to her unpacifiable teeth, until she that crept in by stealth at the mouth, like a puny thief, comes out of the belly like a conqueror through a passage opened by her own labor and industry.

Whether it is true or not that the trochilus awakes the sleeping crocodile when it sees the ichneumon lying in wait to enter the crocodile, I leave to Strabo, who reports it, and to the discretion of the indifferent reader.

Monkeys are haters of crocodiles. The scorpion and the crocodile are also enemies; and therefore, when the Egyptians describe the combat of two notable enemies, they paint a crocodile and a scorpion fighting together.

Dolphins are professed enemies to crocodiles and will fight with them and kill them. There is likewise a certain wild ox or bugil among the Parthians which is an enemy to the crocodile, for if he find or meets with a crocodile out of the water, he is not afraid of him but takes heart and sets upon him and with the weight and violent agitation of his body treads him all to pieces.

Hawks are also enemies to crocodiles. And especially the ibis. If but a feather of the ibis comes upon a crocodile by chance or by the direction of a man's hand, the feather makes the beast immovable and it cannot stir.

There are divers means to take crocodiles. In ancient times they took them with hooks baited with flesh, or else they enclosed them with nets as they do fishes, and now and then with a strong iron instrument cast out of a boat down into the water upon the head of the beast. Leo Africanus relates also this means to take crocodiles. A long, strong rope is tied to a tree upon the banks of the Nile, and at the end of the rope is fastened a hook a cubit long and a finger in quantity. Unto this hook is tied a ram or a goat, which being set close to the river and being tormented by the hook cries out amain. By hearing of this voice, the hunger-greedy crocodile is raised out of his den and invited, as he thinks, to a rich prey, so he comes and swallows the bait, in which he finds a hook not to be digested. Then away he strives to go, but the strength of the rope stays his journey. And to the intent that all of his strength may be spent against the tree and the rope, the hunters are at one end thereof and cause the rope to be cast to and fro, pulling it in, and now letting it go, now terrifying the beast with one noise and fear and anon with another, so long as they perceive in him any spirit of moving or resistance. When he is quiet, they come to him, and with clubs, spears, staves, and such manner of instruments, pierce through the tenderest parts of his body and so destroy him.

Another means of taking crocodiles is said to be the following. Their nature is that, when they go to the land to forage and seek after prey, they cannot return back again but by the same footsteps of their own which they left imprinted in the sand. When the people perceive these footsteps, instantly with all the haste they can make they come with spades and mattocks and make a great ditch, and with boughs cover it so as the serpent may not espy it; and upon the boughs they also again lay sand to avoid all occasion of deceit or suspicion of fraud at his return. Then, when all things are thus prepared, they hunt the crocodile by the foot until they find him. Then with noises of bells, pans, kettles, and suchlike things, they terrify and make him return as fast as fear can make him run toward the waters again, and they follow him as near as they can, until he falls into the ditch, where they come all about him and kill him with such instruments or weapons as they have prepared for him, and so being slain, they carry him to the great city Cairo, where for their reward they receive ten pieces of gold.

There have been some crocodiles brought into that city alive, whereof one was as much as two oxen and two camels could bear and draw, and at the same time was one taken by the device before expressed, which had entered into a village near the Nile and swallowed up alive three young infants sleeping in one cradle. The said infants scarcely dead were taken out of the belly, and soon after, when no more tokens of life appeared, they were all three buried in a better and more proper grave of the earth. Then also there was another crocodile slain, and out of his belly was taken a whole ram not digested, nor any part of him consumed, and a woman's hand which was bitten or torn off from her body above the wrist, for there was upon it a bracelet of brass.

The Indians have a kind of crocodile in the Ganges which has a horn growing out of his nose like a rhinoceros. Unto this beast they cast condemned men to be devoured, for in all of their executions they need not the help of men, seeing they are provided with beasts to do the office of hangmen.

It has been written that Firmus, a tyrant of Egypt, being condemned to the Nile to be devoured by crocodiles, beforehand bought a great quantity of the fat of crocodiles, and so stripping himself stark naked, laid it over his body. So he went among the crocodiles and escaped death, for these savage beasts being deceived with the savor of their own nature spared the man that had but so cunningly carried it.

The flesh of the crocodile is eaten by those people that do not worship it. By the law of God, it is accounted an unclean beast. Yet the taste thereof being found pleasant and the relish good, the common people make use thereof without respect of God or health.

The medicines arising out of the crocodile are many. The first place belongs to the caul, which has more benefits or virtues in it than can be expressed. The blood is held profitable for many things, and among others it is thought to cure the biting of any serpent. Anointing the eyes with the blood cures both the dregs or spots of blood in them and also restores soundness and clearness to the sight, taking away all dullness or deadness from the eyes. And it is said that, if a man takes the liquor which comes from a piece of a crocodile fried and anoints a wound or harmed part, then he shall be immediately rid of all pain and torment. The dung is profitable against the falling off of the hair and many such other things.

The biting of a crocodile is very sharp, deep, and deadly, so that wheresoever he lays his teeth seldom or never follows any cure. But the counsel of physicians is that, as soon as the patient is wounded, he must be brought into a close chamber where there are no windows and there be kept without change of air or admission of light, for the poison of the crocodile works by cold air and light, and by the want of both is cured. But for remedy (if any be) they prescribe the same which is given for the cure of the biting by a mad dog, or (as Avicen writes) the biting of a dog not mad.

The picture of the crocodile was wont to be stamped upon coin, and the skin hung up in many famous cities of the world for the admiration of the people, and there is one at this day at Paris in France.

I close the history of the crocodile with this story. It is reported that the famous grammarian Artemidorus, seeing a crocodile lying upon the sands, was so much touched and moved therewith that he fell into an opinion that his left leg and hand were eaten off by that serpent, and that thereby he lost the remembrance of all his great learning and knowledge of arts.

A Group of Hounds and a Type of Spaniel or Poodle

THE DOG

*T*OPSELL HAS NEARLY *forty-one pages devoted to the dog. A section on dogs in general is followed by a section describing different kinds of dogs. Topsell next includes all of John Caius'* tract, Of English Dogs *(omitted here). The concluding part of the history describes dog diseases and their cures.*

Some parts of the history have a prosaic quality. At times Topsell is overlong, and some passages are difficult to understand. Topsell is at his best in his "dog stories"; the stories about the faithfulness and heroism of dogs are moving. The story of the dog that fought a lion and would not release its hold even in death is one of the best. For the most part, Topsell writes about the excellence of dogs, but he shows that they have also been regarded as base and contemptible creatures.

There is no region or country in the world where dogs are not bred in some store. There are dogs very great: some for hunting, some for war and defense, some for the boar, bull, or bear, some for the hare, cony, or hedgehog. Again, there are smaller dogs: hounds, braches, beagles, shepherds' dogs, house curs, spaniels for both the water and land. And there are dogs for the rich.

In outward appearance, dogs are generally rough. It is a sign of a good constitution if the hair is indifferently long, but if it grows overlong, the mangy scab will follow. The outward proportion of the head alters as the kind of dog alters. The head might be like a lion's or like a hedgehog's. Sometimes the snout might be broad; sometimes it might be piked. A dog's forelegs bend like the arms of a man, and he uses them instead of arms, having five distinct fingers (commonly called claws) upon each foot before, and four upon each foot behind.

The louder and shriller voice of a dog is called barking; the lower and stiller is called whining. To dream of the barking of dogs presages some treasonable harm by enemies, and so likewise if dogs fawn and claw upon a man.

It is the nature of a dog when he makes water to hold up his leg, if he is above six months old or if he has been at procreation. The females do it for the most part sitting, yet some of the generous spirits among females do also hold up the legs. Dogs ever smell to the hinder parts of one another; peradventure thereby they discern their kind and disposition of each other in their own natures. After they have run a course, they relieve themselves by tumbling and rolling to and fro. When they lie down, they turn round in a circle two or three times together, which they do for no other cause but to lie with more ease and from the wind. They sleep as does a man, and they dream very often, as may appear by their often barking in their sleep.

Dogs bear their young the fifth part of a year: that is, about two months and odd days, but this reckoning is not general, for some kinds bear their young three months, and some more. They bring forth many at a time, sometimes five, sometimes seven, nine, or twelve, for so many cells has the female in her womb.

The first they cast forth of their womb is commonly a male, which resembles the father. They are whelped blind and so remain for nine or ten days. It is good to let the whelps suck two months before they are weaned, and that of their own mother, for it is not so good for them to suck another; and in the meantime exercise them to food such as milk, whey, bread, and flesh.

There is much ado to choose the whelp that will prove the best in the litter. Some take for the best the one that sees last. Others remove the whelps from the kennel and lay them apart from one another. Then they watch which of them the bitch takes first and carries into her kennel again, and they take that one for the best. Others make this experiment. First, they compass in the puppies with a little circle of stinking rags and small sticks apt for burning. Then they set them on fire about the whelps, and the puppy which leaps over first, they take for the best, and the one which comes out last, they condemn for the worst.

There is no other creature more loving or more serviceable to his master than a dog, enduring many stripes patiently at the hands of his master and using no other means to pacify his displeasure than humiliation, prostration, and assentation; and after beating, he turns a revenge into a more fervent and hot love.

In their rage, they will set upon strangers. Yet, if any one falls or sits down on the ground and casts away a weapon, they bite him not, taking that declining for submissive pacification.

They meet their master with reverence and joy, crouching or bending a little (like modest persons). And, although they know none but their masters and familiars, yet will they help any man against a wild beast. They remember voices and obey their leader's hissing or whistling.

There was a dog in Venice which had been three years from his master yet knew him again in the marketplace, discerning him from the thousands of people present.

Aelianus thinks that dogs have reason and use logic. A dog in a ship in Africa came to a pitcher of oil to eat some of it, and the mouth of the pot being too narrow for his head to enter in (because the pot was not full), he devised to cast flint stones into the pot, whereby the oil rose to the top and he was able to eat his fill of it. This gives evident testimony that he discerned that heavy things will sink down and that light things will rise up.

There is a nation of people in Ethiopia called Nubae which hold a dog in such admirable estimation that they give unto him the honor of their king, for they have no other king but this dog. If he fawns, they take him for well-pleased; if he barks or flies upon them, they take him for angry; and by his gestures and movings, they conjecture his meaning for the government of their state, giving as ready obedience to his significations as they can to any living, speaking prince of the world.

The Romans and Grecians had a custom to sacrifice a dog in their Lycaean and Lupercal feasts in honor of Pan, who defended their flocks from the wolf; and this was performed in February, yearly, either because dogs were enemies to wolves, or else because dogs drew wolves away from the city by their barking, or else because dogs were reckoned pleasing beasts to Pan, who was the keeper of goats. The Grecians did also offer a dog to the goddess Hecate, who has three heads. One head of a horse, another of a dog, and the third head of a wild man. And the Romans did offer a dog to Genetha for the safe custody and welfare of their household affairs. Their household gods called Lares were pictured and declared to the people sitting in dog-skins, and dogs sitting beside them.

There were dogs sacred in the temple of Aesculapius, because he was nourished by their milk; and Jupiter himself was called *cynegetes* (that is, a Dog-leader), because he taught the Arcadians first of all to hunt away harmful beasts by the help of dogs.

The ancients in many places gave unto dogs solemn funerals in their hallowed cemeteries, and after they were dead, they ceased not to magnify them, as did Alexander, who built a city for the honor of a dog.

All this notwithstanding, many learned and wise men in all ages have reckoned a dog but a base and impudent creature.

Dogs were prohibited and not permitted to enter into the castle of Athens and Isle of Delos because of their public and shameless copulation, and also that no man might be terrified by their presence from supplication in the temples.

Men of impudent wits, shameless behaviors in taking and eating food, were called Cynics. Athenaeus speaks unto Cynics in this sort. "You do not lead abstinent and frugal lives but resemble dogs; this four-footed beast differs from other creatures in four things, and you follow him in his viler and baser qualities: that is, in barking and license of railing, in voracity and nudity."

Stobaeus, in his wicked discourse or dispraise of women, affirms that the curst, sharp, smart, curious, dainty, clamorous, implacable, and wanton-rolling-eyed women were derived from dogs; and Hesiod, to amend the matter, says that when Jupiter had fashioned man out of the earth, he commanded Mercury to infuse into him a canine mind and a clamorous inclination.

But in Chapter 30 of the Proverbs of Solomon, the excellency of a dog is shown: "There be three things which go pleasantly, and a fourth orders his pace aright: the lion, which is the strongest among beasts and fears not the sight of any being; a hunting dog strong in his loins; a he goat; and a king against whom there is no rising up." Now, by this is deciphered a good king, for the lion rises not against beasts unless he is provoked; a dog rises not against his friends but wild beasts; and the he-goat goes before the flock like a guide and keeper.

Now that we have handled the nature of the dog in general, we come to a narration of divers kinds of dogs. And first of all, we begin with the strong and great hunting dogs.

The greyhound or Grecian dog deserves the first place among the hunting dogs, for he can find out his prey reasonably well by scent, is speedy and quick of foot to follow, fierce and strong to take and overcome, and yet silent, coming upon his prey unawares.

These dogs have large bodies, little heads, beaked noses but flat, broad faces above the eyes, long necks (but great next to the body), fiery eyes, and broad backs. Their rage is so great against their prey that sometimes for wrath they lose their eyesight.

The greatest dogs among the strong and great hunting dogs are in India, Scythia, and Hyrcania; and among the Scythians, they join them with asses in yoke for ordinary labor. The dogs of India are conceived by tigers, for the Indians will take divers females or bitches and fasten them to trees in woods where tigers abide. The greedy, ravening tiger comes and instantly devours one or two of them if his lust does not restrain him, and then, being filled with meat, he immediately burns in lust and so limes the living bitches, who are apt to conceive by him. Thus come these valorous dogs, which retain the stomach and courage of their father but the shape and proportion of their mother.

Of this kind were the dogs given to Alexander by the King of Albania when Alexander was going into India and were presented by an Indian whom Alexander admired. Being desirous to try what virtue was contained in so great a body, Alexander caused a boar and a hart to be turned out to the dog, and, when he would not so much as stir at them, he turned bears unto the dog, and he likewise disdained them and rose not from his kennel. The King commanded the heavy and dull beast (for so he termed him) to be hung up. The keeper, the Indian, informed the King that the dog respected not such beasts and that, if he would turn out unto him a lion, he should see what the dog would do.

Immediately, a lion was put unto him, and at the first sight of it, the dog rose with speed (as if never before he saw his match or adversary worthy his strength), and, bristling, the dog made force upon the lion, and likewise the lion at the dog. The dog took the chaps or snout of the lion into his mouth, where he held the lion by main strength until he strangled it. Desirous to save the lion's life, the King willed that the dog should be pulled off, but the labor of men and all their strength were too little to loosen those ireful and deep-biting teeth.

Then the Indian informed the King that, unless some violence was done unto the dog to put him to extreme pain, he would sooner die than let go his hold. Whereupon it was commanded to cut off a piece of the dog's tail, but the dog would not remove his teeth. Then one of his legs was severed from his body, but the dog seemed not appalled. After that, another leg was cut off, and consequently all four. The trunk of his body fell to the ground, but he still held the lion's snout within his mouth. At the last, it was commanded to cut off his head, and after it was done, the bodiless head still hung fast to the lion's jaws. The King was wonderfully moved and sorrowfully repented his rashness in destroying a beast of so noble spirit, which could not be daunted with the presence of the king of beasts: choosing rather to leave his life than depart from the true strength and magnanimity of mind. To mitigate the King's sorrow, the Indian presented unto him four other dogs of the same quantity and nature, by the gift whereof the Indian put away the King's passion and received reward with such recompense as well beseemed the dignity of such a King and also the quality of such a present.

These dogs grow to an exceedingly great stature, and the next unto them are the Albanian dogs. The Arcadian dogs are said to be generated of lions. The dogs of Crete are called *diaponi* and fight with wild boars. The dogs of Epirus are called *chaonides* and are wonderfully great and fierce.

Greyhounds are the least of these kinds, and yet as swift and fierce as any of the rest, refusing no kind of beast except the porcupine, who casts her sharp quills into the mouth of all dogs. The best greyhound has a long body (strong and reasonably great), a neat, sharp head, splendent eyes, a long mouth, sharp teeth, little ears, a straight neck, a broad and strong breast, broad shoulders, round ribs, and fleshy but not fat buttocks. His forelegs are straight and short; and his hind legs, long and straight.

It is the property of greyhounds to be angry with the lesser barking curs, and they will not run after every trifling beast, discerning by secret instinct of nature what kind of beast is worthy or unworthy of their labor, disdaining to meddle with a little or vile creature.

There are in England and Scotland two kinds of hunting dogs found nowhere else in all the world. They call the first kind in Scotland a *rache,* and this is a foot smelling creature of wild beasts, birds, and also fishes which lie hidden among the rocks. The female of this kind is called a *brache* in England.

The second kind is called in Scotland a *sleuth-hound,* being a little greater than the hunting hound and in color for the most part brown or sandy-spotted. The sense of smelling is so quick in these that they can follow the footsteps of thieves and pursue them with violence until they overtake them; and if the thief takes the water, they cast themselves in and swim to the other side, where they find out again afresh their former labor.

The bloodhound differs nothing in quality from the Scottish sleuth-hound except that bloodhounds are greater in quantity and not always one and the same color, for they are sometimes red, sandy, black, white, spotted, or of such color as are other hounds. But most commonly they are brown or red.

The virtue of smelling is in Latin called *sagacitas*. In a dog, it is this sense which searches out and descries the lairs and lodgings of wild beasts. The nature of those dogs that hunt by smelling is, being set on by the voice and words of their leader, to cast about for the sitting of the beast, and so having found it, to follow after it with continual cry till it is wearied, without changing for any other, so that sometimes the hunters themselves take up the beast; at least the hounds seldom fail to kill it. They seldom bark except in the hunting chase; and then they follow their game through woods, thickets, thorns, and other difficult places, being always obedient and attentive to their leader's voice.

The white hounds are said to be the quickest-scented and surest-nosed and therefore best for the hare; the black ones, for the boar; and the red ones, for the hart and roe. But I cannot agree with this, because their color (especially of the latter two) is too like the game they hunt.

The Spanish dogs whom the French call *espagneuls* have long ears, and by their noses hunt both hares and conies. They are not rough but smooth-haired. The Tuscan dogs are commended by Nemesian. Notwithstanding, they are not beautiful to look upon, having deep, shaggy hair. Yet, is their game not unpleasant. The Umbrian dog is sharp-nosed but fearful of his sport. The Aetolian dogs have excellent smelling noses and are not slow or fearful.

The French dogs are derived or propagated of the dogs of Great Britain and are swift and quick-scented, but not all of them, for they have divers kinds.

The grammarians call a dog engendered of a hound and an ordinary French dog *vertagus,* a tumbler.

Other dogs that hunt by smell are terriers or beagles. They will set upon foxes and badgers in the earth and, by biting, expel them out of their dens. Unto these smelling dogs I may also add the water spaniel and the land spaniel.

Dogs that are not of one kind are called mongrels. Among the mongrels are those dogs which the Latins call sociable dogs. These dogs attend upon men in their travels and labors to defend them and are taught to fight for them against men and beasts. There are two sorts of these dogs.

The first sort is lesser, having rough and long, curled hair, the head being covered with long hair. They have a pleasant and tractable disposition, never going far from their masters.

Upon a season a man with his servant and a dog of this sort went on the way to a certain mart to buy merchandise; and, as they traveled, the servant, who carried the purse, diverted a little out of the way to perform the work of nature, and the dog followed him. When the servant was done, he forgot to take up the purse of money that had fallen from him to the ground in that place, and so departed. The dog, seeing the purse, lay down beside it and stirred not a foot. Afterwards, the man and his servant went forward, missing the dog but not their money. When they came to the mart or fair, they were constrained for want of money to return without doing any further thing, resolving to go back again the same way they came to see if they could hear of their money. When at last they came to the place where the servant had left the purse, there they found both dog and money together. The poor cur was scarcely able to see or stand for hunger. When he saw his master and the servant come unto him, he got up from the earth, but life was not able to tarry any longer in his body, and at one and the same time, in the presence of his friends and nourishers, he died. From this it is apparent that one part of the faithful disposition of such dogs is to keep their nourishers' goods committed unto them.

These dogs have fought for their masters and so defended them. Such a dog was the dog of Calvus, who was slain in a certain civil war at Rome; and, his enemies coming about him to cut off his head, his poor dog interposed his body between the blows and would not suffer any foe once to touch his master's carcass, until, by more than six hundred soldiers, the dog was cut in pieces, so living and dying a most faithful companion and thankful friend to the one that had fed him.

The dogs of Gelon, Hieron, Lysimachus, Pyrrhus, Polus, and Theodorus leaped into the burning fires which consumed their masters' dead bodies.

Nicias, a certain hunter, chanced to fall into a heap of burning coals when he was abroad in the woods and, having no help about him but his dogs, there he perished. Yet his dogs ran to the highways and ceased not with barking and apprehending the garments of passersby to show some direful event. At last, one of the travelers followed the dogs and came to the place where the man had been consumed, and by that conjectured the whole story.

Dogs of this sort will try to reveal the murderers of their friends and masters, as the following story shows. As King Pyrrhus traveled in his country, he found a dog keeping watch over the dead body of a man, and he perceived that the dog had almost pined away by tarrying about the body without food. Taking pity on the beast, he caused the body to be buried, and, by giving the dog his belly full of meat, he drew him to love him and so led him away. Afterwards, when Pyrrhus mustered his soldiers and everyone appeared in his presence, the dog also being beside him, the dog saw the murderers of his master, and not containing himself, he set upon them with voice, tooth, and nail. The King, suspecting what had happened, examined them if ever they had seen or known that dog. They denied it, but the King, not satisfied, charged that surely they were the murderers of the dog's master, for the dog had not barked before their appearance but now remained fierce against them. At last, their guilty consciences broke forth at their mouths, and they confessed the whole matter.

The second sort of sociable dogs is greater than the first kind. These greater sociable dogs or defenders are used to guard houses or cattle or are used by soldiers in wars. This kind ought to be horrible and fierce and unacquainted with everyone except his master, so that he will always be at dagger's drawing and ready to fight with all who shall but lay their hands upon him. Art and continual discipline should be used from his littering or infancy to supply in him any defects of nature. Let him be often provoked to wrath by boys, and afterward as he grows, let some stranger set upon him with a weapon, as a staff or a sword, and combat with him till he be wearied. Then let him tear off some piece of the provoker's garment, so that he may depart with a conceit of victory. After the fight, tie him up fast and suffer him not to straggle loose abroad but feed him thus tied up. So shall he in short time prove a strong defender and an eager combatant against all men and beasts that come to deal with him.

Of this sort was the dog of Pheraeus, the tyrant of Thessaly, being a very great and fierce beast and hurtful to all except those who fed him daily. He used to set this dog at his chamber door to watch and guard him when he slept, that whosoever was afraid of the dog might not approach near without exquisite torments.

Nicomedes, King of Bythinia, had one of these great dogs, which he nourished very tenderly and made very familiar with himself. It fell out once upon a time that this King, being in dalliance with his wife Ditizele in the presence of the dog and she hanging about the king's neck and kissing and provoking him to love with amorous gestures, the dog thought that she was offering some violence to his master the King, and he flew upon her and with his teeth pulled her right shoulder from her body and so left the amorous Queen to die in the arms of her loving husband. Which thing caused the King to banish the dog forever out of his sight, and for sorrow whereof he soon afterward died. But the Queen was nobly buried, at Nicomedia in a golden sepulcher: the which was opened in the reign of the Emperor Michael, son of Theophilus; and there the woman's body was found whole and not putrefied, being wrapped in a golden vesture, which taken off and tried in a furnace yielded above a hundred and thirteen pounds of pure gold.

In the next place, we come to the shepherd's dog. There is no creature that will more stir, bark, and make noise than one of these against thief or wild beast. They are used by herdsmen, swineherds, and goatherds to drive away all annoyances from their animals and also to guide and govern the animals upon signs given them by their masters.

The shepherd's dog should be strong, quick, ready, and understanding, both for brawling and fighting, so as he may frighten away and also follow (if need be) the ravening wolf and take away the prey out of his mouth. A square proportion of body is requisite in these beasts, and a tolerable lightness of foot.

The love of these dogs to the animals that they keep is very great, especially to sheep. Publius Aufidius Pontianus bought certain flocks of sheep in the farthest part of Umbria. The Umbrian shepherds with their dogs went along with him to drive the sheep. When the drovers came unto Heraclea and the Metapontine coasts, they left, but the dogs, for love of the sheep, continued and attended them without regard of any man. They foraged in the fields for rats and mice to eat, until at length they grew weary and lean and so returned alone unto Umbria to their masters, who were many days' journey from them.

There is also the village dog or the housekeeper. The village dog ought to be fatter and bigger than the shepherd's dog. He should have a square, strong body and should be great-mouthed, barking bigly. He should be black-colored so that in the night he may seize upon the robber before he discerns the dog's black skin. Therefore, a spotted, branded, parti-colored dog is not approved.

Another sort of dog is the mimic dog, which has a short, recurved body, very long legs, shaggy hair, and a short tail. These dogs are apt at imitating actions or performing strange feats. There is the story of a certain Italian called Andrew who, about the year 1403, had a red dog of strange feats, and this dog was blind. He would be compassed about with a circle of many people in the marketplace. People would bring many rings, jewels, bracelets, and pieces of gold and silver, and there within the circle they were covered with earth. Then was the dog bade to seek them out, and with his nose and feet he did find and discover them at once. Then was he also commanded to give everyone his own ring, jewel, bracelet, or money, which the blind dog did perform directly without stay or doubt.

There is a town in Pachynus, a promontory of Sicily (called Melita) from whence are transported many fine little dogs called *melitaei canes*. They were accounted the jewels of women, but now the said town is possessed by fishermen, and there is no such reckoning made of those tender little dogs. They are not bigger than common ferrets or weasels, yet are they not small in understanding nor mutable in their love to man. They are nourished tenderly for pleasure.

Theodorus, the tumbler and dancer, had one of these, which loved him so well that at his death the dog leaped into the fire after his body.

Nowadays they have found another breed of little dogs besides the Melitaean dogs. They are not above a foot, or half a foot long, and always the lesser the more delicate and precious. Their head is like the head of a mouse but greater; their snout sharp, their ears like the ears of a cony, short legs, little feet, long tail, and white color, and the hairs about the shoulders longer than ordinary are most commended. They are of pleasant disposition and will leap and bite without pinching and bark prettily, and some of them are taught to stand upright, holding up their forelegs like hands; others are taught to fetch and carry in their mouths that which is cast unto them.

There are some wanton women which admit these dogs to their beds, and bring up their young ones in their own bosoms. These dogs are so tender that they seldom have above one young one at a time but they lose their life.

It was reported that, when Grego in Syracuse was to go from home among other gossips, she gave her maid charge of two things. One was that her maid should look to her child when it cried, and the other was that she should keep her little dog withindoors.

Publius had a little dog called Issa, and the dog had about its neck two silver bells upon a silken collar. For the neatness thereof the dog seemed to be a picture rather than a creature.

Having described sundry kinds of dogs, I shall shut up this treatise with a recital of their diseases and the cures.

If you give unto a dog every seventh day or twice in seven days broth or pottage in which ivy is boiled, this will preserve him sound without any other medicine. The small roots of hellebore, which are like to onions, have power in them to purge the belly of dogs. For the drawing forth of a thorn or a splinter out of a dog's foot, take coltsfoot and lard, or the powder thereof burned in a new earthen pot; either of these applied to the foot draws forth the thorn and cures the sore. When a dog becomes deaf, the oil of roses with new pressed wine infused into his ears cures him; and for worms in the ears, make a plaster of a beaten sponge and the white of an egg, and that shall cure him.

When a dog becomes mad, it may be known by these signs. He will neither eat nor drink; he looks awry, and his body is lean. He breathes gaping, and his tongue hangs out of his mouth. His ears are limber and weak; his tail hangs downward. His pace is heavy and sluggish until he runs, and then he is more rash, intemperate, and uncertain, sometimes running, and afterwards standing still again. He is very thirsty but abstains from drink. He barks not and knows no man, biting both strangers and friends.

For the cure of these dogs and first of all for the preventing of madness, there are sundry invented observations. First, it is good to shut them up and make them fast for one day, then purge them with hellebore, and being purged, nourish them with bread of barley-meal. Some say that, if a dog tastes of a woman's milk which she gives by the birth of a boy, he will never fall mad.

When a young male dog suffers madness, shut him up with a bitch; or if a young bitch is also oppressed, shut her up with a male, and the one of them will cure the madness of the other.

If a whelp is cut asunder alive and is laid upon the head of a mad melancholic woman, it shall cure her, and it has the same power against the spleen. The flesh of mad dogs is salted and given in food to them which have been bitten by mad dogs for a singular remedy. The blood is commended against all intoxicating poisons and pains in the small guts, and it cures scabs. The hair of a black dog eases the falling sickness; the brains of a dog in lint and wool laid to a man's broken bones for fourteen days together consolidate and join them together again.

Two Types of Winged Dragons

THE DRAGON

TOPSELL INCLUDES DRAGONS under "serpents." By a serpent, he means a reptile or a creeping or crawling creature. He says that no other serpent is comparable to the dragon and then describes different kinds of dragons: those that have wings and lack feet; those that have wings and feet; and those that lack wings and feet. He describes the dragon as a creature of both good and evil. The stories about dragons that love human beings are especially good.

Among all the kinds of serpents, there is no serpent comparable to the dragon or that affords and yields so much plentiful matter for the ample discovery of its nature. Therefore, I must borrow more time from the histories of some other creatures than peradventure the reader is willing to spare. But such is the necessity that I can omit nothing making to the purpose, either for the nature or morality of this serpent. Therefore, I will strive to make the description pleasant with variable history, seeing that I may not avoid the length hereof, that so the sweetness of the one (if my pen could so express it) may countervail the tediousness of the other.

The Chaldees call a dragon *darkon*, and it seems that the Greek word *drakon* is derived of the Chaldee. Some think that the Greek word may also be derived from *derkein*, because of the vigilant eyesight of dragons, and therefore it is feigned that a dragon had the custody of the Golden Fleece and that dragons have been guardians of many other treasures. The Egyptians did picture their god Serapis with three heads: that is, a lion in the middle; a meek, fawning dog on the right hand; and a ravening wolf on the left hand. All these forms were joined together by the winding body of a dragon turning his head to the right hand of the god. The three heads are interpreted to signify three times: by the lion, the present time; by the wolf, the time past; and by the fawning dog, the time to come. All of which are guarded by the vigilance of the dragon.

There are divers sorts of dragons, distinguished partly by their countries, partly by their quantity and magnitude, and partly by the different form of their external parts. Some dragons have wings and no feet; some have both feet and wings, and some have neither feet nor wings, but are only distinguished from the common sort of serpents by a comb upon their heads and by a beard under their cheeks.

There are dragons of sundry colors, for some are black, some red, some of an ash-color, some yellow. According to the Greek poet Nicander, the shape and outward appearance of dragons are very beautiful.

Some writers, following the authority of Nicander, affirm that a dragon is of a black color, the belly somewhat green. They say that dragons have a treble row of teeth in their mouths upon each jaw and that they have most bright and clear-seeing eyes. They also have two dewlaps of a red color growing under the chin and hanging down like a beard. Their bodies are also set all over with sharp scales, and over their eyes stand certain flexible eyelids. When they gape wide with their mouths and thrust forth their tongues, their teeth seem very much to resemble the teeth of wild swine. And their necks have many times grossly thick hair growing upon them, like the bristles of a wild boar.

The mouth of some dragons (especially of the most tamable dragons) is but little, not much bigger than a pie. But the Indian, Aethiopian, and Phrygian dragons have very wide mouths, through which they often swallow in whole birds and beasts.

Dragons have most excellent senses both of seeing and hearing. The foods of dragons are fruits and herbs or any venomous creature. They live long without food, and when they eat, they are not easily filled. They grow most fat by eating eggs, and in devouring them they use this art. If the dragon is a great one, he swallows eggs whole and then rolls himself, by which he crushes the eggs to pieces in his belly, and so nature casts out the shells and keeps in the edible part. But if the dragon is a young one, as if it be a dragon's whelp, he takes an egg within the spire of his tail and so crushes it hard and holds it fast until his scales open the shell like a knife; then he sucks out of the place opened all the edible part of the egg. In like sort do the young ones pull the feathers off birds which they eat, and the old ones swallow birds whole, casting the feathers out of their bellies.

Because dragons are the greatest serpents, it was wont to be said that, unless a serpent eats a serpent, he shall never be a dragon: for the opinion was that they grew so great by devouring others of their kind. In Aethiopia, they grow to be thirty yards long. Neither have the people any other name for those dragons but "elephant-killers," and they live very long.

The soldiers of Atilius Regulus did kill a dragon which was a hundred and twenty feet long, and the dragons in the dens of the mountain Atlas grow so great that they can scarce move the foreparts of their body.

The greatest dragons of all are in India. Of the Indian dragons, there are two kinds. One kind is fenny, living in marshes. These dragons are slow of pace and lack combs on their heads. The other kind lives in the mountains. They are more sharp and great and have combs upon their heads, their backs being somewhat brown and all their bodies less scaly than those of the other dragons. When they come down from the mountains into the plain to hunt, they are neither afraid of marshes nor violent waters but thrust themselves greedily into all hazards and dangers, and because they have longer and stronger bodies than the dragons of the fens, they beguile them of their food and take away from them their prepared booties. Some of the mountain-dragons are of a yellowish, fiery color, having sharp backs like saws; they also have beards, and when they set up their scales, they shine like silver. The apples of their eyes are precious stones and as bright as fire, in which there is affirmed to be much virtue against many diseases, and therefore they bring unto the hunters and killers of dragons no small gain, besides the profit of their skin and their teeth.

The mountain-dragons commonly have deeper eyelids than the dragons of the fens. Their aspect is very fierce and grim, and whensoever they move upon the earth, their eyes give a sound from their eyelids much like unto the tinkling of brass, and sometimes they boldly venture into the sea and take fishes.

This is the manner in which the Indians kill the mountain-dragons. They take a garment of scarlet and picture upon it a charm in gold letters; this they lay upon the mouth of the dragon's den, for with the red color and the gold, the eyes of a dragon are overcome, and he falls asleep. The Indians in the mean season watch and mutter secretly words of incantation; when they perceive the dragon is fast asleep, suddenly they strike off his neck with an axe, and so take out the balls of his eyes, wherein are lodged those rare and precious stones which contain in them virtues unutterable.

There are Epidaurian dragons, which are bred nowhere but in that country, being tame, and of a yellow golden color, wherefore they were dedicated to Aesculapius.

There are likewise tame dragons in Macedonia, where they are so meek that women feed them and suffer them to suck their breasts like little children. Infants also play with them, riding upon them and pinching them as they would do to dogs without any harm, and sleeping with them in their beds.

The inhabitants of the kingdom of Georgia, once called Media, say that in their valleys there are divers dragons which have both wings and feet and that their feet are like the feet of geese.

In Europe, dragons have also been seen. Even in our own country (by the testimony of sundry writers), divers have been discovered and killed. There have been also dragons many times seen in Germany, flying in the air at midday and signifying great and fearful fires to follow.

When the dragons of Phrygia are hungry, they turn themselves toward the west, and, gaping wide, with the force of their breath they draw into their mouth the birds that fly overhead. It is probable that some vaporous and venomous breath is sent up from the dragon; and the birds, astonished, fall down into the dragon's mouth.

Saint Augustine says that dragons abide in deep caves and hollow places of the earth and that sometimes when they perceive moistness in the air, they come out of their holes, and, beating the air with wings, as it were with the strokes of oars, they forsake the earth and fly aloft. Their wings are of a skinny substance, and very voluble, and spread wide according to the quantity and largeness of the dragon's body.

The bodies of dragons are exceedingly hot, and dragons very seldom come out of the cold earth, except to seek food and nourishment. Because dragons live only in the hottest countries, they commonly make their lodgings near unto the waters or else in the coldest places among the rocks and stones.

Dragons greatly preserve their health by eating wild lettuce, for that makes them vomit and cast forth out of their stomachs any food that offends them; and they are most specially offended by eating apples, for their bodies are much subject to be filled with wind, and therefore they never eat apples but first they eat wild lettuce. Their sight many times grows weak and feeble, and they renew and recover it by rubbing their eyes against fennel or else by eating it.

The age of dragons has never been certainly known, but it is conjectured that they live long and in great health, and therefore they are able to grow so great.

THE DRAGON

There are some people which by certain enchanting verses do tame dragons and ride upon their necks, as a man would ride upon a horse, guiding and governing it with a bridle.

Although dragons are natural enemies to men, like all other serpents, yet many times (if there be any truth in story) they have been possessed with extraordinary love to men, women, and children.

There was one Aleva, a Thessalian neatherd, who did keep oxen in Ossa hard by the fountain Hemonius. There was a dragon that fell in love with this man, for his hair was as yellow as gold, and unto him for his hair did this dragon often come, creeping closely as a lover to his love. And when he came, he would lick his hair and face so gently and in so sweet a manner as the man professed he never felt the like. As the dragon came, so would he go away again, never returning to the man empty but bringing one gift or another, such as his nature and kind could lay hold on.

There was a dragon which loved Pindus, the son of Macedo, King of Emathia. This Pindus had many brothers most wicked and lewd persons, and he being a valiant man of honest disposition, having likewise a comely and goodly personage, and understanding the treachery of his brethren against him, bethought himself how to avoid their hands and tyranny. Now, forasmuch as he knew that the kingdom which he possessed was the only mark they all shot at, he thought it better to leave that to them and so to rid himself from envy, fear, and peril than to imbrue his hand in their blood or to lose his life and kingdom together. Therefore, he renounced and gave over the government and betook himself to hunting, for he was a strong man fit to combat with wild beasts, by whose destruction he made more room for many men upon the earth.

It happened on a day that he was hunting a hind-calf, and, spurring his horse with all his might and main in the eager pursuit of it, he rode out of the sight of all his company, and suddenly the hind-calf leaped into a very deep cave, out of the sight of Pindus, and so saved himself. Then Pindus alighted from his horse and tied him to a tree, seeking out as diligently as he could for a way into the cave. When he had looked a good while about him and could find no way, he heard a voice speaking unto him and forbidding him to touch the hind-calf. This made him look about him to see if he could perceive the person from whom the voice proceeded, but espying none, he grew to be afraid and thought the voice proceeded from some other greater cause, and so leaped upon his horse hastily and departed again to his fellows.

The day after, he returned to the same place, and when he came hither, being terrified with the remembrance of the former voice, he dared not enter into the place but stood there wondering with himself what shepherds or hunters or other men might be in that place to warn him off from his game, and therefore he went round about to seek for some or to learn from whence the voice proceeded. While he was thus seeking, there appeared unto him a dragon of a great stature, creeping upon the greatest part of his body, except his neck and head lifted up a little; and that little was as high as the stature of any man can reach. In this fashion, he made toward Pindus, who, at the first sight, was not a little afraid of him but yet did not run away, but rather gathering his wits together, he remembered that he had about him birds and divers parts of sacrifices, which instantly he gave unto the dragon and so mitigated his fury by these gifts, and as it were with a royal feast, changed the cruel nature of the dragon into kind usage.

The dragon, being smoothed over with these gifts and as it were overtaken with the liberality of Pindus, was contented to forsake the old place of his habitation and to go away with him. Pindus, being no less glad of the company of the dragon, did daily give unto him the greatest part of his hunting as a deserved price and ransom of his life and conquest of such a beast. Neither was he unrequited for it, for Fortune so favored his game that whether he hunted fowls of the air or beasts of the earth he had success and never missed.

His fame for hunting procured him more love and honor than ever could the imperial crown of his country. All young men desired to follow him, admiring his goodly personage and strength, and the virgins and maids fell in love with him and contended among themselves who should marry him. The wives, forsaking their husbands contrary to all womanly modesty, desired his company rather than the society of their husbands. His brethren were enraged against him and sought all means to kill and destroy him. Therefore, they watched for all opportunities, lying in continual ambush where he hunted, to accomplish their accursed enterprise, which at last they obtained. As he followed the game, they enclosed him in a narrow strait near a riverside, where he had no means to avoid their hands; and they and their companions being many, and he alone, they drew out their swords and slew him. When he saw no remedy but death, he cried out aloud for help, and his voice soon came to the ears of the watchful dragon (for no beast hears or sees better).

The dragon came out from his den, and finding the murderers standing about the dead body, he immediately killed them, so revenging Pindus, and then he fell upon the dead body of his friend, never forsaking the custody thereof until the neighbors adjoining the place took knowledge of the fact and came to bury the bodies. But when they came and saw the dragon, they were afraid and dared not come near but stood afar off, consulting what to do. At last, they perceived that the dragon began to take knowledge of their fear. Perceiving their mourning and lamentation for their dead friend and their abstinence from approaching to execute his exequies, the dragon began to think that he might be the cause of their terror. Therefore, he departed, taking his farewell of the body which he loved, and so gave them leave by his absence to bestow upon Pindus an honorable burial, which they performed accordingly. And the river adjoining was named by the name of *Pindus-death*.

There was a dragon, the lover of Aetholis, who came to her every night and did her body no harm, but gently sliding over her, played with her till morning. Then would he depart away as soon as light appeared so that he might not be espied. The maiden's friends came to the knowledge of this and so removed her far away to the intent that the dragon might come no more to her; and thus they remained asunder a great while. The dragon, earnestly seeking for the maiden, wandered far and near to find her. At last, he met her and, not saluting her gently as he was wont, he flew upon her, binding her hands down with the spire of his body, hissing softly in her face, and beating gently with his tail her back parts, as though taking a moderate revenge upon her for the neglect of his love by her long absence.

Another like story is reported of a great dragon which loved a fair woman, beloved also of a fair man. The woman oftentimes did sleep with this dragon, but not so willingly as with the man. Wherefore, she forsook the habitation of her place for a month and went away where the dragon could not find her, thinking that her absence might quench his desire. But he came often to the place where he was wont to meet with the woman, and not finding her, returned quietly back again. At last, he grew suspicious and, like a lover failing in his expectation, grew very sorrowful and so continued till the month was expired, every night visiting the accustomed place. At last, the woman returned, and the dragon presently met with her, and, in an amorous fashion, full of suspicion and jealousy, winding about her body, did beat her as you have heard in the former story.

The examples before expressed are extraordinary, for there is ordinarily hatred between men and dragons. In the discourse of the enemies of dragons, men must have the first place as their most worthy adversaries, for dragons have perished by men, and also men by dragons.

Hercules, when he was a child in his cradle, slew two dragons. And the Corcyraens did worship Diomedes for killing a dragon. Donatus, a holy Bishop in Germany, finding a dragon to lie secretly hidden beside a bridge, killing men, oxen, horses, sheep, and goats, came boldly unto him in the name of Christ, and when the dragon opened his mouth to devour him, the holy Bishop spitting into his mouth killed him.

There was a certain husbandman who, rising early in the morning and traveling, saw by the wayside a great dragon lying still upon the earth without motion, and the man, being weary, thought him to be a trunk of some tree. Therefore, he sat down upon him, and the beast endured him a little while, but at last, he turned his head in anger and swallowed him up.

Vultures, eagles, swans, and dragons are enemies one to another. There is also enmity between dragons and elephants. So great is their hatred that, in Aethiopia, the greatest dragons have no other name but "elephant-killers." Among the Indians also the same hatred remains, and against the elephants the dragons have many subtle inventions. Besides the great length of their bodies, wherewithal they clasp and begirt the body of the elephant, they also continually bite him until he falls down dead, and in the fall they are also bruised to pieces. For the safeguard of themselves, the dragons get and hide themselves in trees, covering their heads and letting the other part hang down like a rope. In those trees, they watch until the elephant comes to eat and crop of the branches; then, before he is aware, they suddenly leap into his face and dig out his eyes. Then do they clasp themselves about his neck and, with their tails or hinder parts, beat and vex the elephant until they have made him breathless, for they strangle him with their foreparts as they beat him with the hinder. In this combat both the elephant and the dragons perish.

Sometimes a multitude of dragons do together observe the paths of elephants, and across those paths they tie together their tails as it were in knots so that when an elephant comes along, they ensnare his legs and suddenly leap up to his eyes, for that is the part they aim at above all others, and they speedily pull them out. If the poor beast delivers himself from present death by his strength, yet he perishes through the blindness received in that combat, for he cannot choose his food by smelling but by his eyesight.

There is no man living that is able to give a sufficient reason of this contrariety in nature between the elephant and the dragon. Many men have labored their wits and strained their inventions to find out the true causes but all in vain except for the one cause that follows. The elephant's blood is said to be the coldest blood of all beasts, and for this cause it is thought by most writers that the dragons in the summertime hide themselves in great plenty in the waters where elephants come to drink. The dragons suddenly leap upon the ears of an elephant, because those places cannot be defended with the trunk, and there they hang fast and suck all the blood out of his body until the poor beast falls down through faintness and dies, and the dragons being drunk with blood do likewise perish in the fall.

The Devil was called a dragon because of his treachery, for he does treacherously set upon men to destroy them.

In the next place, for the conclusion of the history of the dragon, we will take our farewell of him in the recital of his medicinal virtues. First, the fat of a dragon dried in the sun is good against creeping ulcers; and the same mingled with honey and oil helps the dimness of the eyes at the beginning. The head of a dragon keeps one from looking asquint; and if it was set up at gates and doors, it was thought in ancient times to be very fortunate to the sincere worshipers of God. The eyes being kept till they are stale and afterwards being beaten into an oil with honey made into ointment keep anyone that uses it from the terror of night visions and apparitions.

The fat of dragons is of such virtue that it drives away venomous beasts. It is also reported that by the tongue or gall of a dragon boiled in wine men are delivered from the spirits of the night called incubi and succubi, or else nightmares. But above all other parts, the use of the blood is accounted most notable.

I will conclude with the following stories concerning the good success which has been signified unto men and women, either by dreams or sight of dragons.

The mother of the Roman Emperor Alexander Severus dreamed the night before his birth that she brought forth a little dragon. So also did

Olympias, the mother of Alexander the Great, and Pomponia, the mother of Scipio Africanus. The like prodigy gave Augustus hope that he should be Emperor, for, when his mother Aetia came in the nighttime unto the temple of Apollo and had set down her bed or couch in the temple among other matrons, suddenly she fell asleep, and in her sleep she dreamed that a dragon came to her and clasped about her body and so departed without doing her any harm. Afterwards, the print of a dragon remained perpetually upon her belly, so as she never dared any more be seen in any bath.

The Emperor Tiberius Caesar had a dragon which he daily fed with his own hands and nourished like good fortune. At the last, it happened that this dragon was defaced with the biting of emmets and the former beauty of his body much obscured. The Emperor grew greatly amazed thereat, and, demanding a reason thereof of the wise men, he was by them admonished to beware of the insurrection of the common people. And thus with these stories representing good and evil by the dragon, I will take my leave of this good and evil serpent.

The Elephant

THE ELEPHANT

Topsell regarded the elephant as one of the best demonstrations of God's power and wisdom. Its great size was a source of wonder to him, and the many exemplary characteristics possessed by the animal or attributed to it earned his highest praise. The modesty and chastity of elephants, their loyalty to their fellows, their exposure of adulterers and murderers, and their possession of a natural religion among themselves moved Topsell deeply, and he wrote about them with joy.

He says that elephants turn their heads toward the east at the time of copulation and adds that he cannot tell whether they do this "in remembrance of Paradise, or for the Mandragoras, *or for any other reason." The meaning of the passage is complex. Eden was thought to have been in the east. The word* Mandragoras *refers to the mandragora, or mandrake. One belief was that mandragora was a name for the Tree of Knowledge. (See White,* The Bestiary: A Book of Beasts.*)*

There is no creature among all the beasts of the world which is so great and ample demonstration of the power and wisdom of Almighty God as the elephant: both for proportion of body and the disposition of spirit; and it is admirable to behold the industry of our ancient forefathers and their noble desire to benefit us (their posterity) by searching into the qualities of every beast to discover what benefits or harms may come by them to mankind; having never been afraid either of the wildest, but they tamed them; the fiercest, but they ruled them; and the greatest, but they also set upon them. Witness for this the elephant, being like a living mountain in quantity and outward appearance, yet by men so handled as no little dog became more serviceable and tractable.

Elephants are bred in the hot eastern countries. Some writers affirm that the African elephants are much greater than the Indian, but they are in every way inferior to the Indian elephants. For which case, if an African elephant but sees an Indian elephant, he trembles and labors by all means to get out of his sight, as being guilty of his own weakness. Among all elephants, the Indian elephants are greatest, strongest, and tallest, and there are among them two sorts: one greater and the other lesser.

There are elephants in Taprobane and in Sumatra. They are bred in Lybia, in Aethiopia among the Troglodytae, in the mountain Atlas, Syrtes, Zamnes, and Sala, the seven mountains of Tingitania, and in the country of Basman subject to the great Cham.

Of all earthly creatures, an elephant is the greatest. In India, they are nine cubits high and five cubits broad; in Africa, fourteen or fifteen full spans, which is about eleven feet high and proportionable in breadth.

Their color is for the most part mouse-color or black, and there was one all white in Ethiopia. Their skin is so hard and stiff that a sharp sword or iron cannot pierce it. Their head is very great, and the head of a man may easily enter into their mouth as a finger into the mouth of a dog. Their eyes are like the eyes of swine but very red. They have four teeth wherewith they grind their food, and they also have two ivory teeth or tusks which hang forth beyond the other teeth. They keep one of these ivory teeth always sharp to revenge injuries, and with the other one they root up plants and trees for their food.

There is a certain book extant, without the name of the author, written of Judaea or the Holy Land, wherein the author affirms that he saw an elephant's tusk sold to a Venetian merchant for six and thirty ducats, it being fourteen spans long and four spans broad, and it weighed so heavy that he could not move it from the ground.

These ivory teeth have been always of great estimation among all the nations that ever knew them. The Ethiopians paid for tribute unto the King of Persia every third year twenty of these teeth hung about with gold and jet-wood. With the ivory, images and statues for idol gods have been made, as one for Pallas in Athens, one for Aesculapius in Epidaurus, one for Venus under the name of Urania, and one for Apollo at Rome. Solomon had a throne of ivory covered all over with gold, for the costs and charges whereof he could not expend less than thirty thousand talents.

Elephants lose their ivory teeth every tenth year; when they fall off, they bury and cover them in the earth, pressing them down by sitting upon them, and then cover them over with earth by their feet, and so in short time the grass grows upon them: for, when they are hunted, they know it is for no other cause than their tusks, and so when they lose them, they desire to keep them from men, lest the virtues of them being discovered, they should enjoy the less peace and security.

It is admirable what devices the people of India and Africa have invented by natural observation to find out these buried teeth. In the woods or fields where they suspect these teeth to be buried, they bring forth pots or bottles of water and disperse them, here one, there another; and so they let them stand and they tarry to watch them. So, one may sleep, another sing or bestow his time as he pleases. After a little time, they go and look in their pots, and if the teeth lie near their bottles, they draw all the water out of the bottles by an unspeakable and secret attractive power in nature. The watchmen take this for a sure sign and dig about the bottles till they find the teeth; but if the bottles are not emptied, they go to seek in another place.

The trunk of an elephant is a large hollow thing hanging from his nose like skin toward the ground. With his trunk he receives of his keeper whatsoever he gives him. He can take up a small piece of money from the ground with his trunk, and with it he has been seen to pull down the top of a tree which twenty-four men with a rope could not make bend. He fights in war with his trunk, and with it he drives away hunters when he is chased, for he can draw up therein a great quantity of water and shoot it forth again, to the amazement and overthrow of them that persecute him.

Elephants have a wonderful love to their own country, and, although they might be ever so well delighted with divers foods and joys in other places, yet in memory of their country they send forth tears.

They love also waters, rivers, and marshes, so as they are not unfitly called *riparii*, such as live by the rivers' sides. Unless they are sick or watch their young ones, they never live solitary but in great flocks. They live upon the fruits of plants and roots and with their trunks overthrow the tops of trees and eat the boughs and bodies of them. They are so loving to their fellows that they will not eat alone, but having found food, they go and invite the rest to their feasts, more like to reasonable, civil men than unreasonable, brute creatures.

Elephants are delighted above measure with sweet savors, ointments, and smelling flowers, for which cause their keepers will in the summertime lead them into meadows of flowers, and, by the quickness of their smelling, the elephants will themselves choose and gather the sweetest flowers and put them in baskets if the keepers have any. The baskets being filled, the elephants desire to wash themselves, like dainty and neat men, and so they will go and seek water to wash, and then of their own accord they return to the baskets of flowers. If they do not find them, they will bray and call for them. Afterward, having been led into their stable, they will not eat until their keepers take the flowers and dress the brims of their mangers with them and likewise strew their room or standing place with them. Then they please themselves with their food because of the savor of the flowers stuck about their cratch, being like dainty fed persons who set their dishes with green herbs and put them into their cups of wine.

They are most chaste and keep true unto their mates without inconstant love or separation, admitting no adulteries among them. Like men who taste of Venus not for any corporal lust but for desire of heirs and successors in their families, elephants take their venereal complements for the continuation of their kind, without unchaste and unlawful lust. Never above thrice in all their days do the males suffer carnal copulation (the females only twice). Yet is the rage of the males great when the females provoke them, and they do so burn in fury that many times they overthrow trees and houses in India with their tusks and by using their heads like a ram.

They are modest and shamefast about procreation, for at that time they seek woods and secret places, and sometimes the waters because water supports the male in that action, whereby he ascends and descends from the back of the female with more ease. When elephants go to copulation, they turn their heads toward the East, but whether this is done in remembrance of Paradise, or for the Mandragoras, or for any other reason, I cannot tell.

Elephants are never so fierce, violent, or wild but the sight of a ram tames and dismays them, for they fear his horns; and not only a ram, but also the gruntling clamor or cry of hogs. If elephants perceive a mouse run over their food, they will not eat it, for there is in them a great hatred of this creature. As we have shown in the history of the dragon, there is an inbred and native hostility between elephants and dragons. Elephants are also enemies to wild bulls and to the rhinoceros. In the games of Pompey, when an elephant and a rhinoceros were brought together, the rhinoceros ran instantly and whet his horn upon a stone and so prepared himself to fight. He struck most of all at the belly of the elephant, because he knew that it was the tenderest and most penetrable part of the body.

The females are far more strong and courageous than the males, and also they are apt to bear the greater burdens, but in war the male is more graceful and acceptable, because he is taller, giving more assured ensigns of victory. He can carry a wooden tower on his back with thirty men in it and their sufficient food and warlike instruments.

The King of India was wont to go to war with 30,000 elephants of war, and besides these, he also had 3,000 of the strongest elephants in India, which at his command would overthrow trees, houses, walls, or any such thing standing against him.

In the next place, it is good to relate the story of the taking and taming of elephants. In Africa, they take them in great ditches, and when they have fallen therein, the people take them out again with boughs, mattocks, leaves, and the digging down of high raised places, and they turn them into a valley wrought by the labor of men, most firmly walled on both sides, and there with famine they tame them. When an elephant will gently take a bough at the hand of a man, they adjudge him tamed.

The Indians use a more ingenious and speedy means to tame them. First, they dig also a great ditch and place therein such food as the beast loves. Smelling it and coming thereunto, he falls into the ditch for desire of the food. Then a man comes to him with whips and beats him very grievously for a good space, to the great grief of the beast, who through his enclosing can neither run away nor help himself. Then comes another man during this time of punishment and blames the first man for beating the beast. As if afraid of the rebuke, the whipper departs, and the other man pities the beast and strokes him and then goes away. Then comes the whipper again and scourges the elephant as before and that more grievously to the elephant's greater torment for a good space together. The other man comes again and feigns to fight with the whipper and so by force seems to drive him away. This they do successively three or four times. So at the last, the elephant grows to know and love his deliverer, who by that means draws him out and leads him away quietly. When elephants are hurt by a man, they seldom forget and will seek revenge. So the whipper must use a strange and unwonted kind of attire so as he may never be known by the elephant after he is tamed.

The Troglodytae hunt and take elephants in another manner. They climb up into the trees and there sit till the flocks of elephants pass by, and upon the last the watchman suddenly leaps (with great courage), taking hold upon its tail and sliding down to its legs, and with a sharp axe which hangs at his back he cuts the nerves and sinews of the legs with so great celerity that the beast cannot turn about to relieve itself, before it is wounded and made unable to prevent its taking.

There are certain people who eat elephants and are therefore called *Elephantophagi* ("Elephant-eaters"). They do observe the like policy in taking elephants, for by stealth and secretly they set upon the hindmost, or else a solitary elephant, and cut his sinews, which causes the beast to fall down, and then they behead him and eat the hinder parts.

Others among the Troglodytae use an easier, more cunning and less perilous kind of taking elephants. They set on the ground very strong, charged bent-bows, which are kept by many of their strongest young men, and when the flock of elephants passes by, they shoot sharp arrows dipped in the gall of serpents and wound some of them and then follow them by the blood, until they are unable to make resistance.

When elephants are to be tamed, they are fastened to some tree or pillar so as they can neither kick backward nor leap forward, and there hunger, thirst, and famine abate their natural wildness, strength, and hatred of men. Afterward, when their keepers perceive by the dejection of their minds that they are beginning to be mollified and altered, then they give unto them food out of their hands. The beasts begin to cast a far more favorable and cheerful eye upon their keepers, and so at the last necessity frames them unto a contented and tractable course and inclination.

After elephants have been tamed, they grow into civil and familiar uses. They are taught to bend one of their hind legs to take up a rider, who also must receive help from some other persons or else it is impossible to mount on the back of so high a palfrey. They are ruled without bridle or reins, only by a long, crooked piece of wood bending like a sickle and nailed with sharp nails. No man can sit more safely and more softly upon a horse or mule than they do who ride upon elephants.

Elephants are taught many sports, as to dance and leap, and they are apt to learn, remember, meditate, and conceive such things as a man can hardly perform.

Once there was an elephant who played upon cymbals, and others of his fellows danced about him. There was fastened to each of his forelegs a cymbal, and another was fastened to his trunk. The beast would observe just time and strike upon one cymbal, then upon the other, to the admiration of all the beholders.

There was a certain banquet prepared for elephants upon a low bed in a parlor set with divers dishes and pots of wine, whereinto were admitted twelve elephants: six males appareled like men, and six females appareled like women. When they had been admitted, they sat down with great modesty and did not raven upon one dish or the other but partook here and there like discreet, temperate guests, and when they drank, they took the cups and received in the liquor very mannerly, and for sport and festivity they would through their trunks squirt or cast a little of their drink upon their attendants.

The care of elephants to perform the things they are taught is so industrious that, when they are secret and alone, they will practice leaping, dancing, and other strange feats which they could not learn suddenly in the presence of their masters. Pliny affirms for certain truth that an elephant which was dull and hard of understanding was found by his keeper in the night practicing those things which the keeper had taught him with many stripes the day before and which the beast had not learned by reason of his slow conceit.

Elephants are said to discern between kings and common persons, for they adore and bend unto kings, pointing to their crowns.

The King of the Indians was watched by four and twenty elephants, who were taught to forbear sleep and to come in turns at certain hours, and so were they most faithful, careful, and invincible.

Elephants have a kind of religion, for they worship, reverence, and observe the course of the sun, moon, and stars. When the moon shines, they go to the waters wherein she is apparent, and it is observed in Ethiopia that, when the moon is changed until her prime and appearance, these beasts take boughs from off the trees on which they feed and lift them up to heaven, and then look upon the moon, which they do many times together, as it were in supplication to her. In like manner, they reverence the sun rising, holding up their trunk or hand to heaven in congratulation of her rising.

This religion of theirs also appears before their death, for when they feel any mortal wounds or any natural signs of their end, they take up earth, or else some green herb, and lift it up to heaven in token of their innocency and imploration of their own weakness; and in like manner do they when they eat any herb to cure their diseases: first they lift it up to the heavens (as it were to pray for a divine blessing upon it) and then devour it.

I cannot omit their care to bury and cover the dead bodies of their companions or any other of their kind, for finding them dead, they pass not by them till they have lamented their common misery by casting dust and earth on them and also green boughs, in token of sacrifice, holding it execrable to do otherwise.

They have not only an observation of chastity among themselves, but also are revengers of whoredom and adultery in others, as may appear by these examples in history.

A certain elephant, seeing his master absent and another man in bed with his mistress, went unto the bed and slew them both. The like was done at Rome, where an elephant slew both the adulterer and the adulteress and covered them with bedclothes until his keeper returned home, and then by signs he uncovered the adulterers and showed his keeper his bloody tusk that had taken revenge upon them for such a villainy.

And not only thus do elephants deal against the women but they also spare not to revenge the adultery of men. Yea, even of their own keeper. Once there was a rich man who had married a wife not very amiable or lovely, but like himself for wealth, riches, and possessions. After he had gained her possessions, he set his heart to love another more fitting his lustful fancy; and being desirous to marry her, he strangled his ill-favored wife and buried her not far from the elephants' stable and so married the other and brought her home to his house. Abhorring such detestable murder, an elephant brought the new-married wife to the place where the other was buried; and with his tusks he dug up the ground and showed her the naked body of her predecessor, intimating thereby unto her secretly how unworthily she had married a man who was murderer of his former wife.

Their love and concord with all mankind is generally known, especially to their keepers and women. If through wrath they are incensed against their keepers, they kill them, and afterward by way of repentance they consume themselves with mourning.

There is a notable story of an Indian who had brought up from a foal a white elephant, both loving it and being beloved of it. The King hearing of this white elephant sent unto the man for it, requiring it to be given him for a present, whereat the man was much grieved that another man should possess that which he had so tenderly educated and loved. Therefore, like a rival in his elephant's love, he resolved to deny the King and to shift for himself in some other place: whereupon he fled into a desert region with his elephant, and the King understanding thereof grew offended with him and sent messengers after him to take away the elephant and bring the man back again to receive punishment for his contempt.

When the messengers came to the place where the man was, he ascended into a steep place, and there kept the King's messengers off from him by casting of stones, and so also did the beast like one who had received some injury by them. At last, they got near the Indian and cast him down, but the elephant made upon them, killing some of them. He put the rest to flight, and then taking up his master with his trunk, carried him safe into his lodging, which thing is worthy to be remembered as the act of both a loving friend and faithful servant.

The like may be said of the elephant of Porus, who carried his wounded master the King in the battle he fought with Alexander. The beast drew the darts gently out of his master's body without pain and did not cast him until he perceived him to be dead, and without blood and breath. Then did he first of all bend his own body as near the earth as he could so that if his master had any life left in him he might not receive any harm in his alighting or falling down.

At the sight of a beautiful woman, they leave off all rage and grow meek and gentle. There was an elephant in Egypt which was in love with a woman that sold corals. This same woman was wooed by Aristophanes of Byzantium, and therefore it was not likely that she was chosen of the elephant without singular admiration of her beauty. And Aristophanes could say as never a man could that he had an elephant for a rival, and this also did the elephant manifest to the man, for on a day in the market he brought her certain apples and put them into her bosom, holding his trunk a great while therein, handling and playing with her breasts.

Another elephant loved a Syrian woman, with whose whole aspect he was suddenly taken. In admiration of her face, he stroked it with his trunk, in testification of further love. The woman likewise failed not to frame for the elephant amorous devices with beads and corals, silver and such things as are grateful to these brute beasts. So to her great profit, she enjoyed his labor and diligence, and he her love and kindness without all offense to his contentment. At last the woman died, and the elephant, missing her, like a lover distracted between love and sorrow, fell beside himself and so perished.

Elephants observe things done in weight and measure, especially in their food. Once upon a time, an elephant was kept in a great man's house in Syria, having a man appointed to be his overseer, who did daily defraud the beast of his allowance. But, on a day as his master looked on, the overseer brought the whole measure and gave it to the elephant.

Seeing this and remembering how the overseer had served him in times past, the beast exactly divided the corn into two parts in the presence of his master, and so laid one of them aside: by this fact showing the fraud of the servant to his master.

Elephants are most gentle and meek, never fighting or striking man or beast unless they are provoked; but being angered, they will take up a man in their trunk and cast him into the air like an arrow so many times as he is dead before he comes to the ground. Plutarch affirms that, in Rome, a boy pricking the trunk of an elephant with a goad, the beast caught him and lifted him up into the air to shoot him away and kill him; but the people standing by made so great a noise and cry thereat that the beast set the boy down again fair and softly without any harm to him at all, as if he thought it sufficient to have put him in fear of such a death.

In the nighttime, elephants seem to lament with sighs and tears their captivity and bondage, but, if anyone comes, they refrain suddenly like unto modest persons and are ashamed to be found either murmuring or sorrowing.

They live to a long age, even to 200 or 300 years, if sickness or wounds prevent not their life. They are in their best strength of body at threescore, for then begins their youth.

The inhabitants of Taxila in India affirm that they had an elephant at the least three hundred fifty years old, for they said it was the same that fought so faithfully with Alexander for King Porus, for which cause Alexander called him Ajax and did afterward dedicate him to the sun and put golden chains about his tusks with this inscription upon them: "Alexander, the son of Jupiter, consecrates this Ajax to the sun."

I close the history of the elephant with an account of its charity to its own kind. When they wax old and unfit to gather their own food or to fight for themselves, the younger of them feed, nourish, and defend the old ones. Yea, they raise them out of ditches and trenches into which they have fallen, exempting them from all labor and peril and interposing their own bodies for their protection. Neither do they forsake them in sickness or in their wounds but stand to them, pulling darts out of their bodies and helping like skillful surgeons to cure wounds and also like faithful friends to supply their wants.

The Fox

THE FOX

*A*S TO BE EXPECTED, *Topsell emphasizes the cunning and trickery of the fox. But although he describes the fox as a greedy and deceitful animal, he also shows admiration for its cleverness and resourcefulness. Included is much interesting information about methods of hunting foxes.*

The belief that a fox feigned death to trap birds was a traditional one. In the bestiaries this trick is described, and the fox is compared to Satan, who is said to feign death until he has sinners within his grasp.

The epithets expressing the nature of the fox among writers are these: crafty, wary, deceitful, stinking, strong-smelling, quick-smelling, tailed, warlike or contentious, wicked and rough, and (among the Graecians) fiery-colored and subtle for slaughter. Christ called Herod a fox because he understood how, by crafty means, he sought to entrap and kill him.

There are store of foxes in the Alpine regions of Helvetia. And among the Caspians they abound, and their multitude makes them so tame that they come into the cities and attend upon men like tame dogs. The foxes of Sardinia are very ravenous, for they kill the strongest rams and goats and also young calves. In Muscovia, the foxes are black and white. In Spain, they are all white, and their skins are often brought by the merchants to be sold at Francford mart. In the Septentrional or Northern woods, there are black, white, and red foxes, and such as are called "cross-bearing" foxes, for on their backs and overthwart their shoulders there is a black cross, and there are foxes aspersed over with black spots, and all these are of one and the same malignant and crafty nature. It is most commonly seen that foxes which keep and breed toward the South and the West are of an ash color and like to wolves, having loose hanging hair.

Serpents, apes, and foxes, and all other dangerous, crafty beasts have small eyes, but sheep and oxen, which are simple beasts, have very great eyes. When the Germans describe a good horse, they decipher in him the outward parts of many beasts. From the fox they ascribe unto him short ears, a long and bushy tail, an easy and soft treading step, for these belong to the fox. The male fox has a hard, bony genital; his tail is long and hairy at the end; and his temperament and constitution is hot.

The greatest occasion of the hunting of foxes is for the benefit of the skin. The Thracians in the time of Xenophon wore caps of fox-skins upon their heads and ears in the coldest and hardest winters. For this purpose in Germany at this day they slit asunder the skin of foxtails and sew it together again, adding to it a sufficient number till it be framed into a cap. The skin of the belly and sides is of more precious estimation because it is more soft and smooth, and therefore is sold for twice so much as the other parts.

The Muscovians and Tartarians make most account of the black skins, because their princes and great nobles wear them in their garments. The white and blue skins are less esteemed, because the hair falls off and they are lesser than the other. The red skins are most plentiful. Scaliger says that he saw skins brought into France by certain merchants and that these skins had divers white hairs disposed in rows very elegantly upon them, and in divers places they grew also single.

In the summertime, the skins are of little worth, because the beasts are troubled with the falling off or looseness of the hair.

Men who have the gout, shrinking up of the sinews or other fluxions of the rheum in their legs can use no better or more wholesome thing than to wear buskins of the skins of foxes. The Scythians make them shoes and sole them with the backs of the skins of foxes and mice.

The Latins have a proper word for the voice of a fox, which is *gannio, gannire* (to *ganne*), and it is also metaphorically applied to men when by screeching clamors they trouble others.

The abode of foxes in the daytime is in the caves and holes in the earth, and they come not abroad until the night. These dens have many caves in them, and passages in and out, that when terriers set upon foxes in the earth they may go forth some other way. The wolf is an enemy to the fox, and the fox lays in the mouth of his den an herb called sea-onion, which is contrary to the nature of the wolf, and he is so greatly terrified therewith that he will never come near the place where it either grows or lies.

The fox is such a devouring beast that he forsakes nothing fit to be eaten. He kills hares and conies, and with his breath he draws field mice out of their holes and devours them. He devours all kinds of poultry. He also eats grapes, apples, and pears. In Arabia and Syria Palestina, foxes are so ravenous, harmful, and audacious that, in the nighttime, by ganning and barking, they invite one another (as it were) by a watchword to assemble in great multitudes together for to prey upon all things, and they fear not to carry into their dens old shoes and vessels or instruments of husbandry; and for this cause, when the husbandmen hear their barking, they gather all things into their houses and watch them.

But as it falls out that in all gluttonous, ravening persons that while they strive to fill their bellies they poison their lives, so also it fares with foxes, for nature has so ordained that if they eat any food wherein are bitter almonds, they die if they drink not at once.

It is reported that if wild rue be secretly hung under a hen's wing, no fox will meddle with her.

When foxes engender and admit copulation, they are joined like dogs, the male upon the female, and when the female perceives her womb filled, she departs and lives very secret. She brings forth ordinarily four at a time, and those blind and imperfect, without articles in their legs, and they are perfected and framed by licking.

The length of the life of a fox is not certainly known, yet it has been affirmed it is longer than the life of a dog.

If the urine of a fox falls upon grass or other herbs, it dries and kills them, and the earth remains barren ever afterward. The savor of a fox is stronger than any other vulgar beast, for he stinks at nose and tail.

Touching the hunting or taking of foxes, I approve the opinion of Xenophon, who avouches that the hunting of the hare is a more noble game or pastime than the hunting of the fox.

In hunting a fox, the hunter must drive him against the wind, and then he prevents all his crafty and subtle agitations and devices, for it stays his speed in running and also keeps his savor fresh always in the nose of the dogs that follow him. The dogs that kill a fox must be swift, strong, and quick-scented, and it is not good to put on a few at once but a good company together, for be assured that the fox will not lose his own blood till he hazards some of his enemies.

With his tail, which he winds every way, he deludes the hunters. When the dogs are pressed near unto him and are ready to bite him, he strikes his tail between his legs and with his own urine wets it, and so instantly strikes it into the dogs' mouths, whereof when they have tasted, so many of them as it touches will commonly leave off and follow no farther.

Sometimes the fox leaps up into a tree and there stands to be seen and bayed at like a champion in some fort or castle, and although fire be cast at him, yet will he not descend down among the hunters. Yea, he endures to be beaten and pierced with hunters' spears, but at length being compelled to forsake his hold and give over to his enemies, down he leaps, falling upon the barking dogs like a flash of lightning, and where he lays hold he never loosens his teeth or assuages his wrath, till other dogs have torn his limbs and driven breath out of his body.

If a fox takes the earth, then with terrier dogs the hunters ferret him out of his den. In some places, hunters take upon them to take foxes with nets, but this seldom meets with success, because with their teeth they tear them in pieces.

The French have a kind of gin to take a fox by the leg, and I have heard of some who have found the fox's leg in the gin, bitten off with his own teeth from his body, rather putting himself to that torment than to expect the mercy of the hunter, and so he went away upon three feet. Other foxes have counterfeited themselves dead, restraining their breath and not stirring any member when they see the hunter come. Not suspecting any life, the hunter takes the leg of a fox out of the gin, and as soon as the fox perceives himself free, away he goes and never gives thanks for his deliverance. It has been truly said that only the wise and old hunters are fit to take foxes, for they have so many devices to beguile men and deliver themselves that it is hard to know when they are safely taken until they are thoroughly dead.

For the killing of this beast, hunters also use this sleight. They take of bacon grease or bacon as much as one's hand and roast it a little, and therewith anoint their shoe soles, and then they take the liver of a hog cut in pieces, and as they come out of the wood where the beast lodges, they scatter the said pieces in their footsteps and draw the carcass of a dead cat after them. The savor thereof will provoke the fox to follow the footsteps, and then the hunters have a cunning archer or handler of a gun who observes and watches in secret till the beast comes within reach, and then gives him a great and deadly wound.

But if the fox is in the earth and hunters have found the den, then they take this course to work him out. They take a long thing like a beehive which is open at one end and has iron wires at the other end like a grate, and at the open end is set a little door to fall down upon the mouth and to enclose the fox when he enters in by touching of a small rod that supports the door. The frame is set to the mouth of the fox's den, and all other passages are watched and stopped. The fox having a desire to go forth and seeing light by the wires misdeems no harm and enters into the hive, which is wrought close into the mouth of his den, and after he has entered, the rod turns the door fast at the lower end or entrance, and so the fox is entrapped, to be disposed of at the will of the taker.

Foxes are annoyed with many enemies. Among them are the small flies called gnats, against whom the fox uses this policy. He takes a mouthful of straw or soft hay or hair and goes into the water, dipping his hinder parts by little and little. The flies betake themselves to his head, which he keeps out of the water. Then he dips his head underwater to his mouth, wherein he holds the hay as aforesaid, and the flies run thereunto for sanctuary or dry refuge. Perceiving this, the fox suddenly casts it out of his mouth and runs out of the water, and by this means he eases himself of his enemies.

All beasts are his enemies, and he friend and loving to none. With strength, courage, and policy, he deals with every one.

When he finds a nest of wasps or a hive of bees, he lays his tail to the hole and so gathers into it a great many of them, and then he dashes it against a wall or tree or stones adjoining and so destroys them, and thus he continues until he has killed them all and made himself executor to their heaps of honey.

When he perceives a flock of birds in the air, he rolls himself in red earth, making his skin look bloody. Then he lies upon his back, closing his eyes and holding in his breath as if he were dead. The birds, namely crows, ravens, or suchlike, observe this, and because of their hatred of his person, they for joy alight and triumph at his overthrow, and this the fox endures for a good season, till opportunity serving his turn and some of the birds coming near his snout, then suddenly he catches one of them in his mouth and feeds upon it like a living and not a dead fox.

All kinds of hawks are enemies to foxes, and foxes to them, because they live upon carrion. Eagles fight with foxes and kill them. Olaus Magnus affirms that, in the Northern regions, eagles lay eggs and hatch their young in those skins which they have stripped off foxes and other beasts.

Kites, vultures, and wolves are enemies to foxes, because they are all flesh-devouring creatures, but the fox, which has so many enemies, by strength or subtleties overcomes them all.

The flesh of a fox boiled and laid to a sore bitten by a sea hare cures and heals it. A fox-skin (as already said) is profitable against all moist fluxes in the skin, and also the gout, and cold in the sinews. The ashes of a fox's flesh, burnt and drunk in wine, are profitable against shortness of breath and stoppings of the liver. A fox's gall, instilled into the ears with oil, cures the pain in them. The stones of a fox take away pimples and spots in the face. The dung pounded with vinegar, by anointment, cures the leprosy speedily.

Many writers have devised divers witty inventions and fables of foxes to express vices of the world, as when they show a fox in a friar's clothing preaching to hens and geese to signify how irreligious pastors in holy habits beguile the simple with subtlety, also of a fox teaching a hare to say his credo or creed between his legs, and for this cause Almighty God in His Word compares false prophets to foxes, in Ezekiel 13.

The Glutton or "Gulon," the European Wolverine

THE GULON

THE GULON, OR GULO, *lives in North America, Europe, and Asia. The variety found in North America is called the wolverine; in Europe, the animal is called the glutton. The glutton is a fierce animal noted for its voracity. Topsell uses the animal to set forth a vivid lesson: people lose their humanity if they offer "sacrifice to nothing but their bellies."*

This beast was not known by the ancients, but has been since discovered in the Northern parts of the world, and because of its great voracity, it is called *gulo* (a "devourer"). This beast is thought to be engendered by a hyena and a lioness, for in quality it resembles a hyena, and it is the same which is called *crocuta*.

The gulon is a devouring and unprofitable creature, having sharper teeth than other creatures. Some think it is derived of a wolf and a dog, for it is about the bigness of a dog. It has the face of a cat and the body and tail of a fox. It is black of color; its feet and nails are most sharp; its skin is rusty, the hair very sharp.

The gulon feeds upon carcasses. When he has found a carcass, he eats so violently that his belly stands out like a bell. Then he seeks some narrow passage between two trees and there draws through his body, and by pressing it he drives out the food which he has eaten. And being so emptied, he returns and devours as much as he did before, and then goes again and empties himself as in the former manner; and so he continues eating and emptying till all be eaten.

It may be that God has ordained such a creature in those countries where it is found to express the abominable gluttony of the men there, so that they may know their deformed nature. It is the fashion of the noblemen in those parts to sit from noon till midnight eating and drinking, and they never rise from the table but to disgorge their stomachs or ease their bellies, and then they return with refreshed appetites to ingurgitate and consume more of God's creatures: wherein they grow to such a height of beastliness that they lose both sense and reason and know no difference between head and tail. Such are the men in Muscovia, in Lithuania, and most shameful of all in Tartaria.

I would to God that this same (more than beastly, intemperate gluttony) had been circumscribed and confined within the limits of the unchristian or heretical-apostatical-countries and had not spread and infected our more civil and Christian parts of the world. Nobility, amity, good fellowship, neighborhood, and honesty should not ever be placed upon drunken or gluttonous companions, or should any be commended for bibbing and sucking in wine and beer like swine when in the meantime no spark of grace or Christianity appears in them. Such men lose the marks of humanity, reason, memory, and sense; they consume both patrimony and pence in their voracity and forget the badges of Christians, offering sacrifice to nothing but their bellies. The Church forsakes them, the spirit accurses them, the civil world abhors them, the Lord condemns them, the Devil expects them, and the fire of Hell itself is prepared for them and all such devourers of God's good creatures.

There are of these beasts two kinds, distinguished by color. One is black, and the other is like a wolf. They seldom kill a man or any live beasts but live upon carrion and carcasses. Yet sometimes when they are hungry, they prey upon beasts like horses and suchlike. Then they subtly ascend up into a tree, and when they see a beast under it, they leap down upon it and destroy it. A bear is afraid to meet them and unable to match them by reason of their sharp teeth.

This beast is tamed and nourished in the courts of princes for no other cause than for an example of incredible voracity.

When he has filled his belly and can find no trees growing so near together that he can expel his excrements by sliding between them, he goes to an alder tree and rends it asunder with his forefeet and passes through the midst of it for the cause aforesaid.

When these beasts are wild, men kill them with bows and gins for no other cause than for their skins, which are precious and profitable, for the skins are white-spotted, changeably interlined like divers flowers, and for this cause the greatest princes and richest nobles use them in garments in the wintertime, such as the Kings of Polonia, Sweden, and Goatland, and the Princes of Germany. Neither is there a skin which will sooner take a color or more constantly retain it. The outward appearance of the skin is like to a damask garment, and besides this outward part, there is no other memorable thing worthy of observation in this ravenous beast. In Germany it is called a four-footed vulture.

The Hyena

THE HYENA

SOME INFORMATION ABOUT HYENAS *is essential to understand this history. Three kinds of hyenas are recognized today: the spotted hyena* (Crocuta crocuta), *which lives in southern Africa; the striped hyena* (Hyaena hyaena), *found in India, Arabia, and northeast Africa; and the brown hyena* (Hyaena brunnea), *native to southern Africa.*

Topsell, who largely follows Gesner's account of the hyena, correctly identifies one of the three kinds. Beyond that, his description shows confusion and uncertainty. The first kind that he describes is the spotted hyena. The next is not a hyena, but appears to be a baboon. This creature is depicted in an illustration (omitted here) which is that of a baboon or mandrill, and Topsell's description of the creature seems to confirm this conclusion. Topsell says that this kind of "hyena" is called papio or dabuh. Baboons belong to the genus Papio, *but the word* dabuh *is an Arabian name for the striped hyena.*

The third kind of hyena in the history is the crocuta. Topsell says that this beast is an offspring of a male hyena and a lioness. It is uncertain whether he is describing one of the hyenas or another kind of animal. (He also conjectures that the mantichora may be a kind of hyena and discusses it in his history of the hyena, but I have treated it in a separate section.)

In keeping with the customary conception of the hyena, Topsell depicts the animal as hypocritical and treacherous. The hyena, he says, is a great enemy to man, and then he repeats the ancient belief that the hyena can imitate the human voice to lure someone to his death. The history contains most of the traditional lore about the animal.

I find that there are three kinds of these beasts: the common kind of hyena; the kind called *papio* or *dabuh;* and the kind called *crocuta* and *leucrocuta*. It has also been conjectured that the beast called *mantichora* may be a fourth kind.

The first and common kind of hyena is bred in Africa and Arabia, being in quantity of body like a wolf but having rougher hair. He has bristles like a horse's mane all along his back, and in the middle of the back he is a little crooked or dented; his color is yellowish but bespeckled on the sides with blue spots, which make him look more terrible, as if he had so many eyes.

The eyes of the beast change color at his pleasure a thousand times a day, and for this reason many ignorant writers have affirmed the same of the whole body. The hyena cannot see one quarter so perfectly in the day as in the night, and therefore he is called *lupus vespertinus,* a wolf of the night. The skillful lapidarists of Germany affirm that this beast has a stone in his eyes (or rather in his head) called *hyaena* or *hyaenius,* but the ancients say that the apple or pupil of the eye is turned into such a stone and that it is indued with the wondrous quality that if a man lays it under his tongue, he shall be able to foretell and prophesy of things to come. The truth of this I leave to the reporters.

The backbone of the beast stretches itself out to the head, so as the neck cannot bend unless the whole body is turned about. He has a very great heart. The genital member is like a dog's or a wolf's, and I marvel that writers have been possessed with the opinion that these beasts change sexes and are sometimes male and sometimes female (that is to say, male one year and female another).

These beasts engender not only among themselves but also with dogs, lions, tigers, and wolves. The Ethiopian lioness, being covered with a hyena, bears the crocuta, and those beasts called *thoes* are generated between this beast and a wolf.

The hyena is accounted a most subtle and crafty beast. And the female is far more subtle than the male and more seldom taken.

In ancient times it was believed that there was in this beast a magical or enchanting power, for they write that any creature around which a hyena goes three times will stand stone-still and not be able to move out of its place: and if dogs come within the compass of a hyena's shadow and touch it, they immediately lose their voice.

The hyena can counterfeit a man's voice, vomit, cough, and whistle. In the nighttime, she comes to houses or folds where dogs are lodged, and, by making as though she vomited or else whistling, she draws the dogs out-of-doors and devours them. If she finds a man or a dog asleep, she considers whether she or he has the greater body. If she, then she falls on him, and either with her weight or some secret work of nature by stretching her body on him, she kills him or makes him senseless (whereby without resistance she eats off his hands). But if she finds her body to be lesser than his, she takes her heels and flies away.

If a man meets with this beast, he must not set upon it on the right hand but on the left, for it has been often seen that when in haste a hyena did run by the hunter on the right hand, he immediately fell off his horse senseless. Aelianus reports that a hyena coming to a man asleep in a sheep-cote made or cast him into a dead sleep by laying her left hand or forefoot to his mouth. Afterward, she dug about him a hole like a grave, covering all his body over with earth except his throat and head, whereupon she sat until she suffocated and stifled him. Yet Philes attributes such a power to her right foot.

They are great enemies to men, and for this reason Solinus reports of them that by secret accustoming themselves to houses or yards where carpenters or such work, they learn to call their names. Being ahungered, the hyena will come and call one of the men with a distinct and articulate voice, whereby she causes many times the man to forsake his work and go to see the person calling him, but the subtle hyena goes further off and so by calling allures him from help of company, and afterward when she sees the time, she devours him, and for this cause her proper epithet is *aemula vocis* ("voice-counterfeiter").

There is great hatred between a pardal and this beast. If after death their skins are mingled together, the hair falls off from the pardal's skin but not from the hyena's. When the Egyptians describe a superior man overcome by an inferior, they picture these two skins. So greatly are pardals afraid of hyenas that they run away from anything on which any part of a hyena's skin is fastened.

It is said that he who will go safely through the mountains or places of this beast's abode must carry in his hand a root of coloquintida. It is also believed that if a man compasses his ground about with the skin of a crocodile, a hyena, or a sea-calf (a seal) and hangs it up in the gates or gaps thereof, the fruits enclosed shall not be molested with hail or lightning. And for this cause, mariners were wont to cover the tops of their sails with the skins of this beast or of the sea-calf. And it has been said that a man clothed with this skin may pass without fear or danger through the midst of his enemies.

If a man holds the tongue of a hyena in his hand, there is no dog that dares to seize upon him. The skin of the forehead or the blood of this beast resists all kinds of witchcraft and incantation. Pliny writes that the hairs laid to women's lips make women amorous. And so great is the vanity of magicians that they are not ashamed to affirm that if a tooth of the upper jaw on the right side is bound unto a man's arm or any part thereof, he shall never be molested with dart or arrow.

And thus much for the first kind of hyena. Now follows the kind called *papio* or *dabuh*.

Dabuh is the Arabian name of this beast. The Africans call it *Leseph*. In quantity, it resembles a fox; but in wit and disposition, a wolf. Its color is like that of a bear, and therefore I judge it to be *arctocyon*, which is engendered of a bear and a dog. Its feet and legs are like a man's, and it continually holds up its tail, showing the hole behind, for at every motion it turns that as other beasts do their heads. It has a short tail, and but for that, I would judge it to be a kind of ape.

These beasts live two hundred in a company. When they are gathered together, the fashion is for one of them to go before the flock singing or howling, and all the rest answer him with correspondent tune. Their voices are so shrill and sounding that, although they be very remote and far off, yet do men hear them as if they were hard by. When one of them is slain, the rest flock about his carcass, howling as if they made a funeral lamentation.

When these beasts grow to be very hungry because of famine, they enter the graves of men and eat their bodies. Yet is their flesh eaten by men in Syria, Damascus, and Beirut.

They are exceedingly delighted with music, such as by pipes and timbrels. Wherefore when hunters have found out their caves, they spread nets and snares at the mouth thereof, and afterwards they strike up instruments, and the foolish beasts, unaware of any fraud, come out and are taken.

There was one of these beasts in Germany in the year of our Lord 1551 to be seen publicly. It was brought out of the wilderness of India. It ate apples, pears, and other fruits of trees; it climbed up into trees and shook the boughs to make the fruit fall. It was of a cheerful disposition, especially when it saw a woman, whereby it was thought to be a lustful beast. Its four feet were divided like a man's fingers.

I know not whether the dabuh is that kind of little wolf which Bellonius says abounds in Cilicia and Asia and which in the nighttime ravens and comes to the bodies of sleeping men, taking away from them their boots, caps, or bridles.

And thus much shall suffice for the second kind of hyena. The third kind is called *crocuta*. This is not the gulon aforesaid but is a different beast from that. The crocuta is said to be an Ethiopian four-footed beast that is engendered between a lioness and a hyena. Its teeth are all of one bone, being very sharp on both sides of its mouth. They are included in the flesh as in a case, that they might never be dulled. With their teeth they break anything. It is said by Solinus that this beast never winks and that its nature seems to be tempered between a dog and a wolf. Yet it is more fierce than either, more admirable in strength. It also imitates the voices of men to devour them.

So much for the kinds of hyenas. Now follow the medicines arising out of this beast.

The magi or wise men, who have wit in nothing but in circumstances of words, say that the best time to take hyenas is when the moon passes over the sign called Gemini and that for the most part the hairs are to be kept and preserved. The magi do also affirm that the skin of a hyena being spread upon a sore which was bitten by a mad dog does immediately and without any pain cure the sore. The same being bound to a part of the head that does ache will immediately drive away the pain. The same does effectually and speedily help them which are troubled with gout or swelling in the joints. If a hyena's blood being hot is anointed on them which are infested with the leprosy, it will without delay very effectually cure them.

Hyena's flesh being eaten does much avail against the bitings of ravenous dogs, but some are of the opinion that the liver being eaten is of more force and power to cure or heal them.

If one of the great teeth of a hyena is bound with a string unto any that are troubled in the nighttime with shadows and phantasies and are scared out of their sleep with fearful visions, it will very speedily and effectually procure them ease and rest. The tooth of a hyena (called *alzabo*) being bound upon the right arm of anyone who is either oblivious or forgetful, and hanging down from the arm unto the middle finger or wrist, does renew and refresh his decayed memory.

The gall of a bear and of a hyena, being dried and beaten to powder and so mixed with the best honey which is possible to be had and then stirred up and down a long time together, helps those who are stark blind to their eyesight, if it be daily anointed and spread upon the eyes for a reasonable space together.

To conclude, if the dirt or dung which is found in the interior parts of a hyena is burned and dried into powder and so taken in drink, it is very medicinable and curable for those which are grieved with painful excoriations and wringings of the belly and also for those which are troubled with the bloody flux. And the same being mingled with goose grease and anointed over all the body of either man or woman will ease them of any pain or grief which they have upon their bodies whatsoever. The dung itself is also very good to purge and heal rotten wounds and sores which are full of matter and filthy corruption.

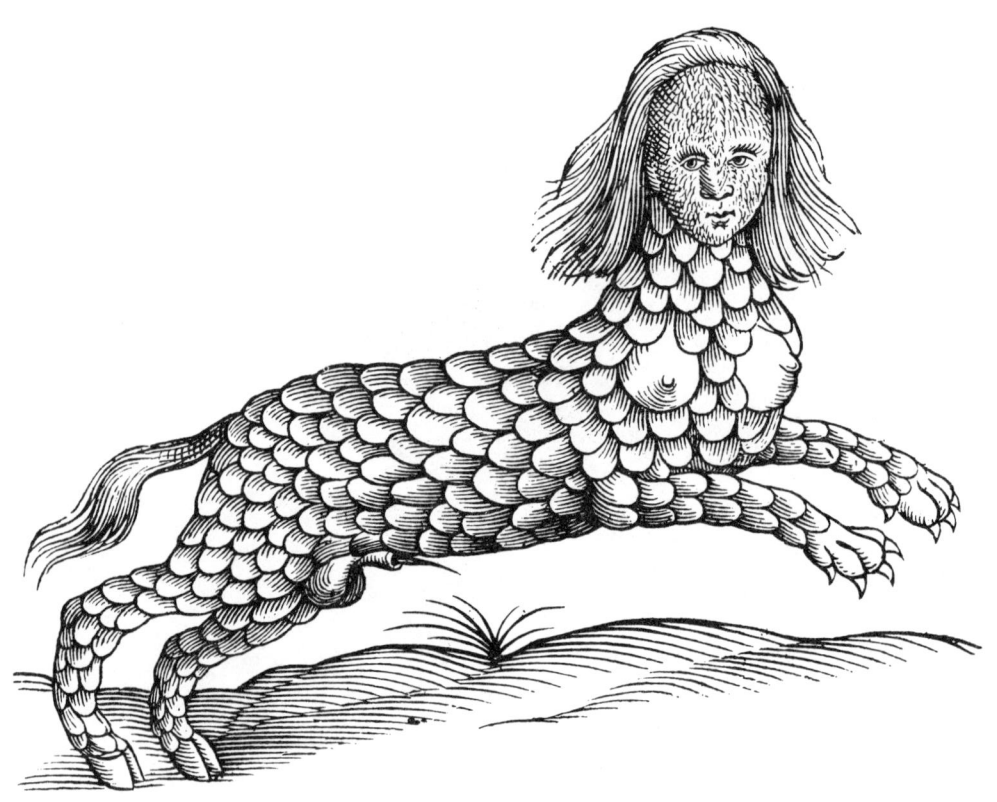

The Lamia

THE LAMIA

IN ANCIENT TIMES, lamiae were said to be monsters that sucked children's blood or preyed upon young men. Some people believed that lamiae could change their appearance and that they usually took alluring shapes to trick victims. A lamia was often described as having a serpent's body and a woman's face and breasts. Topsell thought that such monsters were poetical inventions or were apparitions created by devils. However, he believed that there was a real creature called the lamia and that it was a Lybian beast which had a woman's face and "very large and comely shapes" on its breast.

The word *lamia* has many significations, being taken sometimes for a beast of Lybia, sometimes for a fish, and sometimes for a specter or apparition of women called *phairies.* Some writers have ignorantly affirmed that either there is no such Lybian beast or else that the lamia is a monster compounded of a beast and a fish. In the first place, I shall set down the fables about the lamia, and then I shall come to a description of the actual beast.

When Apollonius and his companions traveled on a bright moonshine night, they saw a certain apparition of phairies (in Latin called *lamiae,* and in Greek, *empusae*) changing themselves from one shape into another, being sometimes visible and presently vanishing out of sight again. As soon as he perceived them, he knew what they were and did rail at them with very contumelious and despiteous words, exhorting his fellows to do the like, for that is the best remedy against the invasion of phairies. And when his companions did likewise rail at them, the vision departed away immediately.

The poets say that Lamia was a beautiful woman which Jupiter loved, bringing her out of Lybia into Italy, where he begot upon her many sons; but Juno, jealous of her husband, destroyed them as soon as they were born, punishing Lamia also with a restless estate that she should never be able to sleep but live night and day in continual mourning, for which occasion she also steals away and kills the children of others. Jupiter, having pity upon her, gave her exemptile eyes that might be taken in and out at her own pleasure, and likewise the power to be transformed into what shape she wished.

Plutarch also affirms that lamiae have exemptile eyes and that as often as they go from home, they put in their eyes, wandering abroad by habitations, streets, and crossways, entering into the assemblies of men, and prying so perfectly into everything that nothing can escape them be it ever so well covered. You will think (Plutarch says) that they have the eyes of kites, for there is no small mote but they espy it, nor any hole so secret but they find it out, and when they come home again, at the very first entrance of their house they pull out their eyes and cast them aside, so being blind at home but seeing abroad. If you ask me (he says) what they do at home, they sit singing and making wool.

There were two women called Macho and Lamo who were both foolish and mad, and from the strange behaviors of them came the first opinion of the phairies. There was also an ancient Lybian woman called Lamia. The opinion was that, if the phairies had not whatsoever they demanded, they would take away live children.

It is reported of Menippus, the Lycian, that he fell in love with a strange woman, who at that time seemed beautiful, tender, and rich, but in truth there was no such thing and all was but a fantastical ostentation.

She was said to insinuate herself into his familiarity after this manner. As he went upon a day alone from Corinth to Cenchreae, he met with a certain phantasm or specter like a beautiful woman, who took him by the hand and told him that she was a Phoenician woman and of a long time had loved him dearly, having sought many occasions to manifest the same but could never find opportunity until that day, wherefore she entreated him to take knowledge of her house, which was in the suburbs of Corinth, therewith pointing unto it with her finger, and so desired his presence. The young man, seeing himself thus wooed by a beautiful woman, was easily overcome by her allurements and did oftentimes frequent her company.

There was a certain wise man, a philosopher, who espied the same and spoke unto Menippus in this manner: "O fair Menippus, beloved of beautiful women, art thou a serpent and dost nourish a serpent?"

By these words, he gave him his first admonition or inkling of a mischief. But this did not prevail with Menippus, for he purposed to marry this specter, her house to the outward show being richly furnished with all manner of household goods.

Then said the wise man again unto Menippus. "This gold, silver, and these ornaments of house are like Tantalus's apples, which are said by Homer to make a fair show but to contain in them no substance at all: even so whatsoever you conceive of these riches, there is no matter or substance in the things which you see, for they are only enchanted images and shadows which you may believe. This your bride is one of the *empusae* called *lamiae,* wonderful desirous of copulation with men and loving their flesh above measure, but those whom they do entice with their venereal marts, they afterwards devour without love or pity, feeding upon their flesh."

At these words, the wise man caused the gold and silver plate and household stuff and cooks and servants to vanish away. Then did the specter like unto one who wept entreat the wise man that he would not torment her or cause her to confess what manner of person she was. But he, being inexorable, compelled her to declare the whole truth, which was that she was a phairy and that she purposed to use the company of Menippus and feed him fat with all manner of pleasures, to the intent that afterward she might eat up and devour his body, for all her kind love was only to feed upon beautiful young men.

These and suchlike stories and opinions in my judgment arise from the prestigious apparitions of Devils, whose delight is to deceive and beguile the minds of men with error contrary to the truth of Holy Scripture, which nowhere makes mention of such enchanting creatures. If there are any such creatures, they are the works of the Devil. Or rather I believe that the lamiae are but allegories devised by the poets: allegories of beautiful harlots who, after they have had their lust of men, do many times devour and make away with them, as we read of Diomede's daughters. And for this reason also harlots are called *lupae* (she-wolves) and *lepores* (hares).

But to leave the fables and come to the true description of the lamia.

In Chapter 34 of Isaiah we find this beast called *lilith* in the Hebrew, and translated by the ancients as *lamia.* So, then, we will take it for granted, by the testimony of Holy Scripture, that there is such a beast as this.

It has been written that there are such beasts in some parts of Lybia, having a woman's face and very beautiful. They have very large and comely shapes on their breasts, and they have no other voice but a hissing like dragons. They are the swiftest of foot of all earthly beasts, so as none can escape them by running.

When a lamia sees a man, she lays open her breasts and by the beauty thereof entices him to come near to conference, and so having him within her compass, she devours and kills him.

It is said that there is a certain place in Lybia, near the seashore, full of sand like to a sandy sea, and all the neighbor places thereunto are deserts. If it happens that through shipwreck men come there on shore, these beasts devour them all. When they see a man, they stand stone-still and stir not until he comes unto them, looking down upon their breasts or to the ground. Some have thought that people are drawn into their compass by a certain natural magical witchcraft. But I cannot approve their opinions. Nor can I believe those writers who describe the lamia as having the feet of a horse and the hinder parts of a serpent. Yet I grant that the lamia kills not only by biting but also by poisoning.

The hinder part of this beast is like unto a goat, her forelegs like a bear's, and her body scaled all over like that of a dragon. Also it is reported of them that they devour their own young ones. And thus much for this beast.

The Lion

THE LION

TO TOPSELL, THE LION *was unquestionably the king of beasts, and he succeeded in making this a history fit for a king. His portrait of the lion is of a fierce, courageous animal with great nobility. Topsell covers nearly the whole body of lore connected with lions. Among the beliefs that were well-known in his time are: a lion will spare a prostrate person; sick lions cure themselves by eating monkeys; lions are terrified by a cock, particularly a white one; and lions repay acts of kindness.*

Being now come to the discourse of the lion (justly styled by all writers the king of beasts), I cannot choose but remember that pretty fable of Aesop concerning the society and honor due unto this beast. For (says he) the lion, ass, and fox entered league and friendship together and foraged abroad to seek convenient booties. At last having found one and taken the same, the lion commanded the ass to make a division of it. The foolish ass regarding nothing but society and friendship and not honor and dignity parted the booty into three equal parts, one for the lion, another for the fox, and the third for himself. The lion, disdaining this because the ass had made him equal to the others, immediately fell upon the ass and tore him to pieces. Then he bade the fox to make the division. The crafty fox divided the prey into two parts, assigning unto the lion almost the whole booty and reserving for himself a very small portion. The lion allowed this and asked him who taught him to make such a partition. "Marry," said the fox, "the calamity of the ass whom you lately tore in pieces."

In like manner, I would be loath to be so simple in sharing out the discourse of the lion as to make it equal with the treatises of the beasts lately handled, but rather according to the dignity thereof, to express the whole nature in a large and copious tractate, for such is the rage of illiterate or else envious men that they would censure me with as great severity, if I should herein like an ass forget myself (if I were in their power), as the lion did his colleague for one foolish partition.

When Lysimachus, the son of Agathocles, was cast by Alexander to a lion to be destroyed, he did not, like a cowardly person, offer himself to the lion's teeth, but he wrapped his arm in his linen garment, and when the lion came gaping at him to devour him, he held him fast by the tongue until he stopped his breath and slew him. For which cause, he was ever afterwards the more loved and honored by Alexander.

In like sort, I will not be afraid to handle the lion and to look into him both dead and alive for the expressing of so much of his nature as I can gather out of any good writer.

According to Aristotle, there are two kinds of lions. The first has a well-compacted, smaller body. They have curled manes, and they are more sluggish and fearful than the lions of the other kind. The body of the second kind is longer; the mane is loose hanging and deeper; and these lions are more noble, generous, and courageous against all kinds of wounds. (And when I speak of manes, it must be remembered that all the male lions are maned but the females are not.)

There are no lions bred in Europe except in one part of Thracia. Yet, in Aristotle's time, there were more famous and valiant lions in that part of Europe lying between the rivers Achelous and Nestus than in all of Africa and Asia.

All the countries in the East and the South lying under the heat of the sun do plentifully breed lions, and except in hot countries, they breed seldom, and therefore the lions of Fez, Temesna, Angad, Hippo, and Tunis are accounted the most noble and audacious lions of Africa, because they are hot countries. The lions of colder countries have not half so much strength, stomach, and courage. The Lybian lions have not half so bright hair as others; their face and neck are very horribly rough, making them look fearsome, and the whole color of their bodies is between brown and black. Apollonius saw lions also beyond the Nile, the Hyphasis, and the Ganges, and Strabo affirms that there are lions about Meroe, Astave, and Astaboras, which lions are very gentle, tame, and fearful.

Ethiopia also breeds lions, being black-colored, having great heads, long hair, rough feet, fiery eyes, and mouths between red and yellow. Cilicia, Armenia, and Parthia, about the mouth of the Ister, breed many fearsome lions, having great heads, thick and rough necks and cheeks, bright eyes, and eyelids hanging down to their noses. There are also plenty of lions in Arabia, so that a man cannot travel near the city Aden over the mountains with any security of life unless he has a hundred men in his company.

The lions of Hyrcania are very bold and hurtful; and India, the mother of all kinds of beasts, has most black, fierce, and cruel lions. In

Tartaria, also, and the kingdom of Narsinga, and the province of Abasia, are many lions, greater than those of Babylon and Syria, of divers and sundry intermingled colors: white, black, and red. There are many lions in the province of Gingui, and for fear of them men dare not sleep out of their own houses in the nighttime, for whomsoever they find, they devour and tear in pieces. The ships which go up and down the river are not tied to the bankside for fear of these lions, because, in the nighttime, they come down to the waterside, and if they can find any passage into the barks, they enter in and destroy every living creature. Therefore, the ships ride at anchor in the middle of the river.

The color of lions is generally yellow. The hair of some of them is curled, and some of them have long, shaggy, thin hair, not standing upright but falling flat, longer before and shorter behind; and although the curling of the hair is a token of sluggish timidity, yet if the hair is long and curled at the top only, this portends abundant animosity. So also if the hair is hard, for beasts that have soft hair, as the hart, the hare, and the sheep, are timorous, but those which are harder haired, as the boar and the lion, are more audacious and fearless.

A lion has a most valiant and strong head. The face is not round as some have imagined (therefore comparing it unto the sun, because, in the compass thereof, the hairs stand out eminent like sunbeams), but rather it is square-figured. The eyes are red, fiery, and hollow, not very round nor long, looking for the most part awry. The pupils or apples of the eye shine exceedingly, insomuch as beholding them a man would think he looked upon fire. The mouth is very great, gaping wide; the lips are thin, so that the upper parts fall in the nether, which is a token of fortitude.

A lion's shoulders and breasts are very strong, as also the forepart of his body, but the members of the hinder part do degenerate. His claws are crooked, and exceedingly hard, and it seems a little miracle in nature that leopards, tigers, panthers, and lions do hide their claws within their skin when they go or run, so that they might not be dulled, and never pull them forth except when they are to take or devour their prey: also, when they are hunted, with their tails they cover their footsteps with earth so that they may not be revealed.

The voice of the lion is called *mugitus* (that is, roaring or bellowing). Their sight is most excellent. They sleep with their eyes open, and because of the brightness of their eyes, they cannot endure the light of fire: for fire and fire cannot agree. Also, their smelling is very eminent, and for this cause they are tamed in Tartaria and used for hunting boars, bears, hares, roebucks, wild asses, as also for wild and outlandish oxen, and they were wont to be carried to hunting: two lions in a cart together, with a little dog following each lion.

There is no beast more vehement than a she or female lion, for which cause Semiramis, the Babylonian tyranness, esteemed not the slaughter of a male lion or a libbard, but having gotten a lioness, she rejoiced therein above all others.

When a lion eats, he is most fierce, and also when he is hungry. But when he is satisfied and filled, he lays aside that savage quality and shows himself of a more meek and gentle nature, so it is less danger to meet with him filled than hungry, for he never devours any till famine forces him.

I have heard a story of an Englishman in Barbary who had turned Moor and lived in the King's court. On a day it was said in his presence that there was a lion within a little space of the court, and the place was named where it lodged. The Englishman, being more than half drunk, offered to go and kill the lion hand to hand, and he armed himself with a musket, sword, and dagger, and other complements, and he had also about him a long knife. So forth went this renegade English Moor, more like a madman than an advised champion, to kill this lion; and when he came to it, he found it asleep, so that with no peril he might have killed her with his musket before she saw him; but he, like a foolhardy fellow, thought it as little honor to kill a lion sleeping as a stout champion does to strike his enemy behind the back. Therefore, with his musket top, he smote the lion to awake it, whereat the beast suddenly mounted up and without any thanks or warning set her forefeet on the squire's breast and with the force of her body overthrew the champion, and so stood upon him, keeping him down, holding her grim face and bloody teeth over his face and eyes, a sight, no doubt, that made him wish himself a thousand miles from her, because, to all likelihood, those teeth should be the grinders of his flesh and bones, and his first executioners to send his cursed soul to the Devil for denying Jesus Christ his Savior.

Yet it fell out otherwise, for the lion having been lately filled with some liberal prey did not begin to eat him but stood upon him for her own safeguard and meant so to stand till she was hungry, during which time, the poor wretch had liberty to gather his wits together, and so at the last, seeing that he could have no benefit by his musket, sword, or dagger and perceiving nothing before him but unavoidable death, he thought for the saving of his credit, so that he might not die in foolish infamy, to do some exploit upon the lion whatsoever did betide him. Thereupon seeing the lion did bestride him, standing over his upper parts, and having his hands at some liberty, he drew out his long knife and thrust it twice or thrice into the lion's flank. The lion endured this, never hurting the man, but supposing the wounds came some other way, and thus she did not forsake her booty to look about for the means whereby she was harmed. At last finding herself sick, her bowels being cut asunder within (for in all hot bodies wounds work immediately), she departed away from the man above some two yards distant, and there lay down and died. The wretch being thus delivered from the jaws of death, you must think made no small brags in the court, notwithstanding he was more beholding to the good nature of the lion, which does not kill to eat unless it be hungry, than to his own wit, strength, or valor.

The male lion does not feed with the female, but they eat apart by themselves. Whatsoever lions leave of their meat, they return not to it again to eat it afterwards. Some have assigned the cause of this to be in the meat, because lions can endure nothing unsweet, stale, or stinking; but in my opinion, they do it through the pride of their natures, resembling in all things a princely majesty, and therefore scorn to have one dish twice presented to their own table.

Tame lions being driven by hunger will eat dead bodies, and they will also eat cakes made of meal and honey, as may appear by the tame lion which came to Apollonius and was said to have the soul in it of Amasis, King of Egypt. The story is related by Philostratus in this manner.

There was (says he) a certain man which by a leash led up and down a tame lion like a dog whithersoever he would, and the lion was not only gentle to his leader but to all other persons. By this means, the man got much gain and visited many regions and cities, not sparing to enter into the temples at the time of sacrificing. The lion did not lick up the blood of the beasts nor once touched the flesh cut in pieces for the holy altar but did eat upon cakes made with meal and honey, also bread, gourds, and boiled flesh, and now and then at customary times did drink wine. As Apollonius sat in a temple, the lion came to him in more humble manner than ever he did to any, lying down at his feet and looking up in his face, as if he had some special supplication unto him, and the people thought he did it for hope of some reward, at the command and for the gain of his master.

At last Apollonius looked upon the lion and told the people that the lion did entreat him to signify unto them what he was: namely, that he had in him the soul of a man, that is to say, of Amasis, King of Egypt, who reigned in the province of Saïs, at which words, the lion sighed deeply and mourned forth a lamentable roaring, gnashing his teeth together and crying with abundance of tears. Whereat, Apollonius stroked the beast and made much of him, telling the people that his opinion was, forasmuch as the soul of a king had entered into such a kingly beast, he judged it altogether unfit that the beast should go about and beg his living, and therefore they should do well to send him to Leontopolis, there to be nourished in the temple. The Egyptians agreed and made sacrifice to Amasis, adorning the beast with chains, bracelets, and branches, so sending him to inner Egypt, the priests singing before him all the way their idolatrous hymns and anthems.

Of the transfiguration of men into lions, we shall say more afterward. I rehearsed this story in this place only to show the food of tame and enclosed lions.

Wherefore as lions excel in strength and courage, so also they do in cruelty, devouring both men and beasts, setting on troops of horsemen, depopulating flocks and herds of animals, carrying some alive to their young ones, killing five or six at one time, and whatsoever they lay hold on, they carry it away in their mouth, although it be as big as a camel.

They love camel's flesh exceedingly. When Xeres led his army through Paeonia over the river Chidorus, lions came and devoured his camels in the nighttime. They neither meddled with the men, oxen, nor victuals but only the camels. So it seems no meat is so acceptable unto them.

They hate above measure wild asses and hunt and kill them. They eat apes, but more for physic than nourishment. They also eat young elephants.

It is said that when the beasts hear the voice of the lion all of them do keep their standing and dare not stir a foot. This assertion wants not good reason, for by terror and dread they stand amazed.

Notwithstanding the great valiancy of lions, they have their terrors, enemies, and calamities. We have shown already in the story of dogs that the great dogs in India and Hyrcania do kill lions and forsake other beasts to combat with them. There is also a tiger called *lauzani,* which, in many places, is twice as big as a lion, that kills them. A serpent does easily kill a lion.

The cock is a terror to the lion and the basilisk, and the lion runs away when he sees a cock, particularly a white cock. The reason that the lion fears the cock is that they are both partakers of the sun's qualities in a high degree, and therefore the greater body fears the lesser because there is a more eminent and predominant sunny property in the cock than in the lion.

There is a beast called *leontophonus,* a little creature in Syria, and if the lion tastes its flesh, he dies.

The noises of wheels and chariots terrify lions, and they are afraid of a little, strange noise. A man traveling in the mountains was, by reason of frost, cold, and snow, driven into a lion's den, and at night when the lion returned, he scared him away by the sound of a bell. The like shall also be declared of wolves in their story.

Lions are also afraid of fire, for, as they are inwardly filled with natural fire (for which cause by the Egyptians they were dedicated to Vulcan), so are they the more afraid of outward fire.

There is no beast more desirous of copulation than a lioness, and for this cause the males oftentimes fight. Sometimes eight, ten, or twelve males follow one lioness, like so many dogs one bitch in heat.

The lioness commits adultery by lying with the libbard or the leopard, and if she does not wash herself before she comes to her mate, he discovers her adultery and punishes her. When she is ready to be delivered of her adulterous offspring, she flies to the lodgings of the libbards and there among them hides her young ones (which for the most part are males), for if the male lion finds them, he knows them and destroys them as a bastard issue. When she goes to give them suck, she feigns as though she went to hunting.

Some say that a lioness bears only once in all her life and then only one whelp. But experience shows the contrary. The truth is that a lion bears never above three times: six at the first, no more than two the next time, and lastly but one. Some are of the opinion that the whelps are brought forth without life and that they remain lifeless for three days, until by the roaring of their father and by breathing in their face they are quickened. But Isidorus declares that for three days and three nights after their littering they do nothing but sleep and at last are awakened by the roaring of their father. So it seems without controversy they are senseless for a certain space after their whelping.

There is no creature that loves her young ones better than the lioness, for both shepherds and hunters frequenting the mountains do oftentimes see how irefully she fights in defense of her young, receiving wounds from many darts that open her bleeding body and from many stones that press the blood out of the wounds, standing invincible, never yielding until death. Yea, death itself is nothing unto her so that her young might never be taken out of her den.

In Pangaeus, a mountain of Thracia, there was a lioness which had whelps in her den, the which den was observed by a bear. On a day finding the den unfortified, both by the absence of the lion and the lioness, the bear entered and slew the lions' whelps, and afterward went away and, fearing revenge, climbed up into a tree for her better security against the lions' rage, and there sat as in a sure castle of defense.

At length, the lion and the lioness returned home, and finding their little ones dead in their own blood, they both according to natural affection fell exceedingly sorrowful to see them so slaughtered whom they both loved. But smelling out the murderer by foot, they followed with rage up and down until they came to the tree where the bear was; and seeing her, they assayed to get into the tree but in vain. So the tree hindered them from revenge and gave unto them further occasion of mourning and unto the bear occasion to rejoice at her own cruelty and to deride their sorrow.

Then the male forsook the female, leaving her to watch the tree, and he like a mournful father for the loss of his children wandered up and down the mountain making great moan and sorrow, till at the last he saw a carpenter hewing wood, who seeing the lion coming toward him let fall his axe for fear, but the lion came very lovingly towards him, fawning gently upon his breast with his forefeet and licking his face with his tongue; which gentleness of the lion the man perceiving, he was much astonished, and being more and more embraced and fawned on by the lion, he followed the beast, leaving his axe behind him, but the lion went back and made signs with his foot that the carpenter should take it up. Perceiving that the man did not understand his signs, the lion brought it himself in his mouth and delivered it unto him, and so led him into his cave, where the young whelps lay all imbrued in their own blood, and then led him where the lioness did watch the bear. She seeing them both coming, as one that knew her husband's purpose, did signify unto the man that he should consider the miserable slaughter of her young whelps, and showed him by signs that he should look up into the tree where the bear was. When the man saw the bear, he conjectured that the bear had done some grievous injury unto them. Therefore, he took his axe and hewed down the tree. The bear tumbled down headlong, and the two furious beasts tore her all to pieces. And afterwards, the lion conducted the man unto the place and work where he had first met him, and there left him without doing the least violence or harm unto him.

Neither do the old lions love their young ones in vain and without thanks or recompense, for in their old age they requite it again. Then do the young ones defend them from the annoyances of enemies and also maintain and feed them by their own labor, for they take them forth to hunting, and when in their decrepit and withered estate they are not able to follow the game, the younger pursue and take it for them. Having obtained it, they roar mightily to signify unto their elders that they should come to dinner, and if they delay, they go to seek them where they left them, or else the younger ones take it to them. At the sight whereof, in gratulation of natural kindness and also for joy of good success, the old ones lick and kiss the younger, and afterward they enjoy the booty in common.

The clemency of lions is worthy of commendation and to be wondered at in beasts, for if one prostrates himself unto them as if in petition for his life, they often spare him, except in extremity of famine. Likewise, they seldom destroy women or children, and if they see women, children, and men together, they take the men who are strongest and refuse the others as weaklings and unworthy their honor.

If they happen to be harmed by a dart or stone by any man, they frame their revenge according to the quality of the hurt, for if they are not wounded, they only terrify the hunter; but if they are pinched further and have blood drawn, they increase their punishment.

In Lybia, the people believe that lions understand the petitions and entreatings of them that speak to them for their lives. There was a certain captive woman coming home again into Getulia, her native country, and when she was set upon by many lions, she used no other weapon against them but threatenings and fair words, falling down on her knees unto them, beseeching them to spare her life, telling them that she was a stranger, a captive, a wanderer, a weak, a lean and lost woman and therefore not worthy to be devoured by such courageous and generous beasts. And at these words they spared her, which thing she confessed after her safe return. The name of this woman was Juba.

Although some question whether it is true that the lion will spare a prostrate suppliant making confession unto him that he is overcome, yet the Romans did so generally believe it that they caused to be inscribed so much upon the gates of the great Roman Palace.

The disposition of lions is admirable, both in their courage, society, and love, for they love their nourishers and other men with whom they are conversant.

When Androcles, a servant of a Senator of Rome, had run away because he had committed some offense, he went into Africa and obtained another master. Because his new master whipped him, he was compelled to run away again. He went into the solitary places of the fields and the sands of the wilderness, and he was so scorched with the heat of the sun that he found a cave and did cover himself from the heat therein; and this cave was a lion's den. After the lion had returned from hunting (being very much pained by reason of a thorn which was fastened in the bottom of his foot), he uttered forth such great lamentation and pitiful roarings, by reason of his wound, that it seemed as though he wanted somebody unto whom he could make complaint for remedy. Coming to his cave and finding a young man hidden there, he gently looked upon him and began as it were to flatter him and offered him his foot and did as well as he could to get him to pull out the thorn. But the man at the first was very sore afraid of him and made no other reckoning but of death; but after he saw such a huge, savage beast so meek and gentle, he began to think that surely there was some sore on the beast because he lifted up his foot so unto him, and then taking courage, Androcles lifted up the lion's foot and found in the bottom of it a great thorn, which he plucked forth, and so by that means eased the lion of his pain. He pressed forth the matter which was in the wound, and without any great fear he carefully and thoroughly dried the wound and wiped away the blood. Being eased of his pain, the lion lay down to rest, putting his foot into the hands of Androcles.

The lion was so very well pleased with his cure that he not only spared his life, but he also went daily abroad to forage and brought home the fattest of his prey. All this while (even for the space of three years), the lion kept Androcles familiarly in his den without any note of cruelty or evil nature, and the man and the beast lived mutually at one commons, the man roasting his meat in the hot sun and the lion eating his part raw.

When Androcles had thus lived for three years and had grown weary of such habitation, life, and society, he bethought himself of some means to depart; and therefore, when the lion was gone abroad to hunt, the man took his journey away from that hospitality; and after he had traveled three days (wandering up and down), he was apprehended by legionary soldiers, to whom he told his long life and habitation with the lion and how he had run away from his master, a Senator of Rome.

He was sent to Rome to the Senator, and he was guilty of so great and foul faults that he was condemned to death. And the manner of his death was to be torn into pieces by wild beasts. Now there were at Rome in those days many great, fearsome, cruel, and ravening beasts, and among them many lions. It happened that, shortly after the taking of the man, the aforesaid Lybian lion, seeking abroad for his companion and man-friend, was taken and brought to Rome and put among the other lions. He was the most fierce, grim, fearsome, and savage in the whole company, and the eyes of men were more fastened upon him than upon any of the others. When Androcles was brought forth to his execution and cast among the savage beasts, this lion at the first sight looked steadfastly at him, stood still a little, and then came toward him softly and gently, smelling to him like a dog and wagging his tail. The poor man, not looking for anything but present death, trembled and was scarce able to stand upright in the presence of such a beast, not once thinking upon the lion that had nourished him so long. Mindful of former friendship, the beast licked gently his hands and legs and so went round about him touching his body, and so the man began to know him, and both of them to show joy about their meeting. The man stroked and kissed the lion, and the lion fell down prostrate at the man's feet.

A pardal came with open mouth to devour the man, but the lion rose up against her and defended his old friend and tore her to pieces, to the great astonishment of the beholders. Then Caesar sent for the man and asked him the cause of that so rare and prodigious event, and the man told him the story before expressed. The story was quickly spread abroad among the people, and tables of writing were made of the whole matter, and finally all men agreed that it was fit that both the man and the lion should be pardoned and restored to liberty. So this was done, and the lion was given unto Androcles, who led him up and down the streets by a leash. Androcles received money, and the lion was adorned with flowers and garlands, and all men that saw or met them said,

"Here goes the lion which was this man's host, and here is the man who was this lion's physician."

In the next place, I shall speak of the fictions of metamorphosing and transfiguring men into lions, which was promised in the former discourse of Amasis and Apollonius when I discoursed of the food of lions.

Timaeus the Pythagorean affirms that the mutation of men into beasts is but a fiction brought in for the terror of wicked men who cannot be restrained from vice by the love of well-doing but may be deterred by the fear of punishment which is meant by such beastly transfigurations.

He that lives by catching and snatching to serve his own concupiscence is said to be transmuted into a kite; he that has a love of military discipline and martial affairs, into a lion; he that was a tyrant and a devourer of men, into a dragon. Empedocles said that, if a man departs this life and is transmuted into a brute beast, it is happiest for him if his soul goes into a lion; but if he is transmuted into a plant, then it is best to be metamorphosed into a laurel or bay tree. And for these causes we read of Hippo changed into a lion and Atlas into a lioness, and the like I might say of Proteus and others.

Generally, all the Eastern wise men believed the transmigration of spirits from one into another. Therefore, they taught that all their priests after death were turned into lions; their religious vestals or women, into hyenas; their servants or ministers in the temples, into crows and ravens; and the fathers of families, into eagles and hawks.

Aelianus reports that, when lions in Lybia are deceived in their hunting and cannot get prey to satisfy their hunger, they go to the houses of men, and if they find the man at home, they will enter in and destroy unless they are resisted by wit, policy, and strength. But if they find no man but only women, the women drive them away by rebukes and railing at them, which thing shows their understanding of the Lybian tongue.

The sum and manner of the speeches and words which the women use to affright and turn them away from entering houses are these:

> Art not thou ashamed being a lion, the king of beasts, to come to my poor cottage to beg meat at the hands of a woman? And like a sick man distressed with the weakness of body to fall into the hands of a woman, that thou by her mercy mayst attain those things which are requisite for thy own maintenance and sustentation? Yea, rather thou shouldst keep

in the mountains and live in them by hunting the hart and other beasts, provided in nature for the lion's food, and not after the fashion of little base dogs come and live in houses to take meat at the hands of men and women.

By such words, the women enchant the mind of the lion; and notwithstanding his own want, hunger, and extremity, he casts his eyes to the ground ashamed and afflicted like a reasonable person overcome with strong arguments, and departs away without any enterprise.

Neither ought any judicious or wise man think this thing to be incredible, for we see that horses and dogs which live among men and hear their continual voices do discern also their terms of threatening, chiding, and rating and so stand in awe of them. Therefore, the lions of Lybia, whereof many are brought up like dogs in houses, with whom the little children play, may well come to the knowledge and understanding of the Mauritanian tongue.

Lions are also said to have understanding of the parts of men and women and to discern sexes and to be inued with a natural modesty, declining the sight of women's privy parts.

When the tail of the lion stands immovable, it shows that he is pleasant, gentle, meek, unmoved, and apt to endure anything, which falls out very seldom, for in the sight of men he is seldom found without rage. In his anger, the lion first of all beats the earth with his tail, afterwards his own sides, and lastly he leaps upon his prey or adversary.

There are diverse ways of hunting and taking lions. The Indian dogs and some other strong dogs are used to hunt bulls, boars, and lions. In some places, people make pitfalls or great ditches in the ground to catch lions. In other places hunters do set upon lions with firebrands and drive them into nets where they are taken.

Lions which are tamed in their infancy are most meek and gentle, full of sport and play, especially being filled with meat. When lions grow to be old, it is impossible to make them utterly tame. Yet, we read in divers stories of tame lions, whether made so from littering or else constrained by the art of man.

Hanno had a certain lion which in his expeditions of war carried his baggage, and for that cause the Carthaginians condemned him to banishment, for they said that it was not safe to trust such a man with the government of the commonwealth, who by wit, policy, or strength was able to overcome and utterly to alter the wild nature of a lion, for they thought that he would prove a tyrant who could bring the lion to such meekness as to wait on him at table, to lick his face with his tongue, and to live in his presence like a little dog.

The Indians tame lions and elephants and set them to plough. Onomarchus, the tyrant of Catana, had lions with which he did ordinarily converse.

When lions are tamed, they should be handled neither too roughly nor too mildly, for if they be beaten with stripes they grow overstubborn; and if they be kept in continual flatteries and used overkindly, they grow overproud.

The best way to tame lions is to bring up with them a little dog and oftentimes to beat the dog in their presence, by which discipline the lion is made more tractable to the will of his keeper.

Having spoken of the taming of lions, it now follows to treat of the length of their life and the diseases incident unto them, with their cures. It is held that lions live very long, as threescore or fourscore years, for it has been seen that, when a lion has been taken alive, and in his taking received some wound by which he became lame, or lost some of his teeth, yet did he live many years; and also it is found that some lions have been taken without teeth, which had all fallen out of their heads through age.

The sicknesses with which they are annoyed are not very many, but those they have are continual. For the most part, their entrails or inner parts are never sound but are subject to corruption. Also, by reason of their extremely hot nature, they suffer one sickness or another.

In his best estate, even when the lion seems to be in good health, he is afflicted with a quartan ague, and if this disease did not restrain his violence and malice by weakening his body, he would be far more hurtful to mankind than he is.

When he feels himself sick through the abundance of meat, he falls a vomiting, or else he eases himself by eating a kind of grass, or green corn in the blade, or else rapes; and if none of these prevails, he fasts and eats no more till he finds ease; or else, if he can meet with an ape, he devours and eats its flesh, and this is his principal remedy and medicine which he receives against all his diseases both in youth and age; and when he grows old, being no more able to hunt harts, boars, and such beasts, he exercises his whole strength in hunting and taking apes, whereupon he lives totally.

Unto this natural discourse of lions belongs the use of their parts.

The ancient Moors and barbarians between the mountain Caucasus and the river Cophen clothed themselves in the skins of lions as well as the skins of panthers, and also they slept upon their skins in the night. Hercules is pictured wearing a lion's skin that the world might be admonished what was the ancient attire of our forefathers. Adrastus was commanded by the Oracle to marry his daughters to a boar and a lion when they came a wooing them. When Tydeus came in a boar's skin and Polynices in a lion's skin, Adrastus gave them his daughters in marriage, taking it to be the meaning of the Oracle that men clothed in those skins should be the husbands of his daughters.

It is said that a man anointed all over with the blood of a lion shall never be destroyed by any wild beast. The flesh of a lion being eaten by a person who is troubled with dreams and fantasies in the nighttime will very speedily and effectually work him ease and quietness. And being taken in drink, the flesh helps those who are troubled with the palsy or the shaking of the joints.

Whosoever shall have shoes made of the hide or skin of a lion or a wolf and wear them upon his feet, he shall never have any pain or ache in them. The skin or hide of a lion is also very good for either a man or woman troubled with the piles or swelling of the veins if they shall but at some several times sit upon it.

The fat of a lion is reported to be contrary to poison and venomous drinks, and being taken in wine, it will by the scent expel all wild beasts from anyone. The grease, being mixed with oil of roses, eases and helps those who are troubled constantly with agues and quartan fevers.

If the liver of a lion is dried and beaten to powder and put in the purest wine that can be gotten and is so drunk, it takes away the pain and grief from any person who is troubled with his liver.

Unto this discourse, I shall add pictures and images of lions both in temples and upon shields. In the temple where the shield of Agamemnon was hung up, the picture of Fear was drawn with a lion's head, for the lion sleeps little and in his sleep his eyes are open, and when he wakes he shuts them. So is the condition of Fear. Therefore, the ancients did symbolically picture a lion upon the doors of their Temples; and also they engraved the figure of lions upon ships.

It was also a usual custom to picture lions about fountains and conduits, especially among the Egyptians, that the water might spring forth of their mouths. The river Alpheus was called the *Lion's fountain* because, at the heads thereof, there were dedicated the pictures of many lions.

At the entrance of Thermopylae upon the tomb of Leonidas, the captain of the Spartans, there stood a lion of stone. Upon the steps of the Capitol of Rome, there were two lions of black marble touchstone. King Solomon built his ivory throne upon two lions of brass; and upon the steps of stairs ascending up to that throne were placed twelve lions, here and there. And from hence it came that many kings and states gave in their arms the lion rampant, passant, and regardant, distinguished in divers colors in the fields of or, argent, azure, and sables, with such other terms of art.

There is a constellation in heaven called the Lion, and he is the greatest and most notable sign among the signs of the Zodiac, containing three stars in his head and one clear one in his breast. When the sun comes to that sign, the vehement heat of the sun burns the earth and dries up the rivers. And because the lion is also of a hot nature and seems to partake of the substance and quantity of the sun, he has that place in the heavens. In heat and force, the lion excels all other beasts, as the sun does all other stars.

In his breast and foreparts, the lion is most strong, and in his hinder part more weak. So is the sun, increasing until the noon or forepart of the year, until the summer, and afterwards it seems to languish towards the setting or later part of the year, called the winter. The lion is also a signification of the sun, for the hairs of his mane do resemble the streaming beams of the sun.

The people called Ampraciotae did worship a lioness because she killed a tyrant. The Egyptians built a city to the honor of lions, calling it Leontopolis and dedicating temples to Vulcan for their honor. And in the porches of Heliopolis, there were common stipends for nourishing lions.

The chimera is feigned to be compounded of a lion, a goat, and a dragon. The mantichora has the body of a lion, and the leucrocuta or crocuta has a neck, tail, and breast like a lion's.

It is reported that Meles, the first King of Sardis, did beget of his concubine a lion, and the soothsayers told him that on what side soever of the city he should lead that lion, it should remain inexpugnable and never be taken by any man. Whereupon, Meles led him about every tower and rampier of the city which he thought was weakest, except only one tower standing towards the river Tmolus, because he thought that side was invincible and could never by any force be entered, scaled, or ruinated. Afterwards, in the reign of Croesus, the city was taken in that place by Cyrus.

To conclude, we find in Holy Scripture that our Savior Christ is called the lion of the tribe of Judah, for, as he is a lamb in his innocency, so is he a lion in his fortitude. And the Devil is called a roaring lion, because lions in their hunger are most of all full of fury and wrath.

The Manticore or Mantichora

THE MANTICHORA

THE MANTICHORA, or *manticore*, is a fabulous beast; and a real animal (possibly the lion or the tiger) might have been the basis for this composite creature. Writers on natural history were uncertain about what kind of animal the mantichora was. Topsell conjectures that it might be a kind of hyena and describes it in his history of the hyena. For the sake of clarity, I have treated the mantichora in a separate section.

Citing a description given by Ctesias of Cnidus (fl. 400 B.C.), Topsell says that the mantichora has a body and feet like those of a lion, a face like that of a man, a treble row of teeth in each jaw, and a tail tipped with sharp, pointed quills. He points out that the mantichora is the same beast that is called leucrocuta; but his description of the leucrocuta differs somewhat from that of the mantichora.

Some writers have thought that the mantichora may be a kind of hyena. This beast or rather monster (as Ctesias writes) is bred among the Indians, having a treble row of teeth beneath and above. Its greatness, roughness, and feet are like a lion's; its face and ears are like a man's; its eyes are gray; and its color, red. Its tail is like the tail of a scorpion, armed with a sting, and casting forth sharp, pointed quills. Its voice is like the voice of a small trumpet or pipe. It is as swift as a hart. Its wildness is such that it can never be tamed, and its appetite is especially to the flesh of man. Its body is like the body of a lion, being very apt both to leap and to run, so as no distance or place hinders it.

I take it to be the same beast which Avicen calls *marion* and *maricomorion*. With her tail, she wounds her hunters, whether they come before her or behind her, and immediately when the quills are cast forth, new ones grow up in their place, wherewithal she overcomes all the hunters. Although India is full of divers ravening beasts, yet none of them are styled with a title of *anthropophagi* (that is to say "men-eaters") except only the mantichora.

When the Indians take a whelp of this beast, they bruise its buttocks and tail so that it may never be fit to bring sharp quills. Afterwards, it is tamed without peril.

This is also the same beast which is called *leucrocuta:* about the bigness of a wild ass, being in legs and hooves like a hart, having its mouth reaching on both sides to its ears, and the head and face of a female like a badger's. It is also called *martiora,* which in the Persian tongue signifies a "devourer of men."

The Panther or Leopard, or "Pardal, Libbard"

THE PANTHER,

COMMONLY CALLED A PARDAL, A LEOPARD, AND A LIBBARD

As Topsell points out, there was great uncertainty whether the panther, the pardal, and the leopard were different beasts or were one kind of beast produced by crossbreeding. Topsell accepts the latter explanation and concludes that panthers are the largest of these creatures, pardals are the next in size, and leopards are the smallest.

Topsell portrays the panther as a beautiful but deceitful animal. The belief that the panther attracted other animals by its sweet smell was one of the most familiar beliefs about the creature. Aristotle, Pliny, and Aelian are among the writers who mention the belief.

There have been so many names devised for this one beast that it has grown a difficult thing either to make a good reconciliation of the authors who are wed to their several opinions or else to define the beast perfectly and make of him a good methodical history. Seeing that the greatest variance has arisen from words, I trust it shall be a good labor to collect out of every writer that which is most probable concerning this beast and express the best definition thereof we can learn out of all.

To begin with, all the question has arisen from the Greek and Latin names, and thus it is most requisite to express them and show how the different construction began. The Grecians use *pordalis, pardalis,* and *panther;* the Latins, *panthera, pardalis, pardus,* and *leopardus.* These names are distinguished thus by the learned: *pordalis* signifies the male; and *pardalis,* the female; *panthera* among the Latins is used for the female, and *pardus* for the male. These beasts are understood to be of a simple kind without commixture of generation. *Leopardus* (the leopard or libbard) is a word devised by later writers upon the opinion that this beast is generated between a pardal and a lion, and so it should be properly taken, if there be any such.

Pliny is of the opinion that *pardus* differs from *panthera* in nothing but in sex, and others say that between the lions and the pardals there is such a confused mixed generation as is between asses and mares or stallions and asses: for example, when the lion covers the pardal, then is the whelp called *leopardus,* a leopard or libbard, but when the pardal covers the lioness, then it is called *panthera,* a panther.

Until somebody can show me better, I will subscribe to the opinion that all of these creatures are but one kind of beast and differ in quantity only through adulterous generation. The greatest they call *panthers.* The second they call *pardals,* and the third, the least of all, they call *leopards.*

In Africa, there is great want of waters, and therefore the lions, panthers, and other beasts do assemble themselves in great numbers together at the running rivers, and there the pardals and the lions do engender one with another: I mean the greater panthers with the lionesses, and the greater lions with the panthers; and so likewise the smaller with the smaller, and thereby it comes to pass that some of them are spotted and some without spots.

The pardal is divers colored and very swift, a fierce and cruel beast, very violent, having a body and mind like ravening birds, and some say that pardals are engendered now and then between dogs and panthers, or between leopards and dogs. It is the nature of these pardals in Africa to get up into the rough and thick trees, where they hide themselves among the boughs and leaves, and they not only take birds but also from thence leap down upon beasts and men when they espy their advantage.

Concerning the leopard, the word itself is new and lately invented. The leopard is bred between the pardal and the lioness and is like a lion in the head and form of his members, but he is lesser and nothing so strong.

From the sight of a leopard's skin, Gesner made this description: "The length from the head to the tail was as much as a man's stature and half a cubit. The tail itself three spans and a half, the breadth in the middle three spans, the color a bright yellow distinguished into divers spots, the hair short and mossy."

The leopard is a wrathful and angry beast, and whensoever it is sick, it thirsts after the blood of a wild cat and recovers by sucking that blood or else by eating the dung of a man. Above all other things, it delights in the camphor tree, and therefore lies underneath it to keep it from spoil.

The leopard is sometimes tamed and used instead of a dog for hunting, both among the Tartarians and other princes, for they carry them behind them on horseback, and when they see a deer or hart or convenient prey, they turn them down upon them suddenly, and they take them and destroy them. Yet such is the nature of this beast as also of the pardal that if he does not take his prey at the fourth or fifth jump he falls so angry and fierce that he destroys whomsoever he meets. Yea, many times the hunter. Therefore the hunters have always a regard to carry with them a lamb or a kid or some such living thing, wherewithal they pacify him after he has missed his game, for without blood he will never be appeased.

The Greek word *pardalis* or *pordalis* (for they signify one) seems to me in most probability to be derived from the Hebrew word *pardes,* signifying a garden, because as colors in a garden make it spotted and render a fragrant smell, so the panther is divers colored like a garden of sundry flowers, and also he is said to carry with him a most sweet savor wherever he goes.

The teeth of the panther are like saws, and the tongue is of such incredible sharpness that, in licking, it grates like a file.

The female panther is oftener taken than the male, because she is enforced to seek abroad for her own meat and that of her young ones. The place of the abode of panthers is among the mountains and woods, and especially they delight in the tree camphor. They raven upon the flesh of both birds and beasts. For this cause, they hide themselves in trees, especially in Mauritania, where they are not very swift of foot, and therefore they give themselves to take apes, which they attain by this policy.

When the panther sees the apes, she makes after them, who at her first approaching climb up into the tops of trees, and there sit to avoid the panther's teeth, for she is not able to follow them so high. But she is more cunning than the apes and devises more shifts to take them.

She goes under the tree where the apes are lodged and lies down as though she were dead, stretching out her limbs, restraining her breath, shutting her eyes, and showing all other tokens of expiration. The apes that sit on the tops of the trees behold from on high the behavior of their adversary, and because all of them wish her dead, they more easily believe that which so much they desire. Yet they dare not descend to make trial.

Then, to end their doubts, they choose out one from among them all who they think is of the best courage, and him they send down as it were for a spy. Forth then he goes with a thousand fears in his mind and leaps from bough to bough with no haste (for dread of an ill bargain). Yet being come down, he dares not approach nigh, but having taken a view of the counterfeit and repressed his own fear, he returns again. After a little space, he descends the second time and comes nearer the panther than before, yet returns without touching her. Then he descends the third time, looking into her eyes. He makes trial whether she draws breath or not, but she keeps both breath and limbs immovable, by that means emboldening the apes to their own destruction.

The spy-ape sits down beside the panther and stirs not. Now, when those above in the trees see how their intelligencer abides constantly beside their adversary without harm, they gather their spirits together and descend down in great multitudes, running about the panther, first of all going upon her and afterwards leaping with great joy and exultation, mocking this their adversary with all their apish toys, and testifying their joy for her supposed death. And in this sort, the panther suffers them to continue a great season, till she perceives they are thoroughly wearied, and then upon a sudden she leaps up alive again, taking some of them in her claws, destroying and killing them with teeth and nails, till she has prepared for herself a rich dinner out of her adversaries' flesh. As Ulysses endured all the contumelies and reproaches both of his maids and his wife's suitors until he had a just occasion given him for revenge, so does the panther the disdainful dealings of the apes.

So great is the love of these beasts to all spices and aromatical trees that they come over all the mountain Taurus through Armenia and Cilicia when the winds bring the savor of the sweet gum to them, out of Pamphilia from the tree storax: whereupon lies this story. There was a certain panther which was taken by King Arsaces and a golden collar put upon his neck, with this inscription, "King Arsaces to the God Bacchus." This beast grew very tame and would suffer himself to be handled and stroked by the hands of men, until the springtime that he winded the savor of the aromatical trees and then he would run away from all his acquaintance, and so at last he was taken in the nether part of the mountain Taurus, which was many hundred miles distant from the King's court of Armenia.

Aelianus says that the panther or pardal smells most sweet, which savor he has received from a divine gift. When he feels himself to be hungry and stands in need of meat, then does he get up into some rough tree, and by his savor or sweet smell, he draws unto him an innumerable company of wild goats, harts, roes, and hinds, and such other beasts, and upon a sudden leaps down upon them when he espies his convenient time.

Solinus says that the sweetness of the panther's savor works the same effect upon beasts in the open fields, for they are so mightily delighted with his spotted skin and fragrant smell that they will always come running unto him from all parts, striving who shall come nearest him to be satisfied with the sight, but when once they look upon his fierce and grim face, they are all terrified and run away. For this cause, the subtle beast turns away his head and keeps that from their sight, offering the more beautiful parts of his body, as an alluring bait to a mouse, and so he destroys them.

Albertus is of the opinion that the report of the panther's savor or sweet smell is but a fable, because (he says) it is written as a maxim among philosophers that no creatures (man excepted) can be said to smell either sweet or sour.

I dissent from Albertus about the savor of the panther, and my reason is this. It is granted by all men that the beasts do assemble about the panther because of either the spots of his skin or his sweet savor. Therefore it follows that if he is lodged in a tree and not visible to the eyes of the beasts and they assemble about him in the tree where he is lodged, there can be no cause for their assembling but the attractive power of his sweet savor.

Therefore I will conclude this point with admiration of the work of the Creator, to consider how wisely He has disposed His goodness and how powerfully He communicates the affections of His divinity even unto brute beasts, for He does not distinguish them asunder only by their outsides and exterior parts or by their insides and the qualities of their minds but also by the air they draw in and the savor they send forth.

At the time of their lust, panthers have very peculiar voices. At the sound of the voices, other beasts come about them, as lions, lionesses, wolves, and thoes.

They never bear above once, because, when the young ones begin to stir in the dam's belly and gather strength for birth, they cannot tarry the just time of their delivery but tear out with the sharpness of their nails the womb or bag wherein they lie, and therefore their dam is forced for the avoiding of pain to cast them forth of the womb both blind and deformed. Afterwards, she can never conceive again by reason that her womb is so torn with the claws of her first whelps that it is not able to retain to perfection the received seed of the male.

Panthers live together in flocks and greatly delight in their own kind but in no other kind as far as I know.

As a lion does in most things imitate and resemble a man, so does a panther a woman, for it is a fraudulent, though a beautiful, beast. The disposition of a panther is wanton, effeminate, outrageous, treacherous, deceitful, fearful, and yet bold. And for this reason in Holy Scripture it is joined with the lion and the wolf to make the triplicity of ravening beasts.

Their love to their young ones is exceedingly great, for if at any time while they are abroad to forage they meet with hunters that would take their young away, they fight for them unto death; and to save them from blows they interpose their own bodies, receiving mortal wounds. If they find their young ones taken out of their den in their absence, they bewail their loss with loud and miserable howling.

There is the story of a panther that lay in the highway to meet a man to help her young ones out of a ditch or deep pit in which they had fallen. At length, there appeared in her sight the father of the philosopher Philinus, who immediately began to run away as soon as he saw the beast, but the poor distressed beast rolled after him in a humble manner as though she had some suit unto him and took him lightly by the skirt of his garment with one of her claws. Perceiving that she gave suck by the greatness of her udders hanging under her belly, he began to take pity upon her and laid away fear, thinking that indeed her young ones had been taken away from her by one means or another. Therefore he followed her, she drawing him with one of her feet unto the place where her young ones had fallen, and he delivered them to the mother as a ransom for his own life, and then both she and the young ones did follow him rejoicing, dismissing him without all manner of harm out of the danger of all beasts and out of the wilderness. It is a rare thing in a man to be so thankful, and much more in a beast.

There was once a man who had brought up a tame panther from a whelp and had made it so gentle that it refused no society of men, and he himself loved it as if it had been his wife. There was also a little kid in the house brought up tame, of purpose to be given unto the panther when the kid had grown to some stature or quantity. In the mean season, the panther played with it every day. At last, it being ripe, the master killed it and laid it before the panther to be eaten, but it would not touch it, whereupon it fasted till the next day, and then the kid was brought unto it again, but it refused it as before. At last, it fasted the third day, and making great moan for meat, it had the kid laid before it the third time. The poor beast seeing that nothing would serve the turn but that it must either eat up its chamber fellow or else its master would make it continually fast, it ran and killed another kid, disdaining to meddle with that which was its former acquaintance, yea though it were dead. Herein it excelled many wicked men who do not spare those that have lived with them in the greatest familiarity and friendship to undo and overthrow them alive for the advancement of themselves.

The great panther is a terror to the dragon, and so soon as the dragon sees it, he flies to his cave. There is great hatred and enmity between the hyena and the panther, for in the presence of the hyena, the pardal dares not to resist; and what is more admirable is, if there is a piece of a hyena's skin about either man or beast, the panther will never touch it.

Leopards are afraid of a tree called *Leopard's tree.* Panthers are also afraid of the skull of a dead man and run from the sight of it. It is reported that, two years before the death of Francis, the King of France, two leopards, a male and a female, escaped in France into the woods, and they tore in pieces many men and women. At last, they came and killed a woman who was to have been married that day, and afterward there were found many carcasses of women destroyed by them, of which they had eaten nothing but their breasts.

When panthers are hunted and are forced into the presence of hunters, they leap directly unto their heads. Therefore, a hunter takes great care both of his standing and also of holding his spear, for if he receives not the panther in his leap and does not gore him to the heart or else otherwise wound him mortally, he is gone and his life is at an end.

In hunting of wild beasts, the wary woodman must make good choice of his horse, not only for the mettle and agility which are very necessary, but also for the color. For the gray horse is fittest for the bear and most terrible to him. The yellow or fire color against the boar, but the brown and reddish color against the panther.

The Moors use this device to take panthers and all such harmful beasts. They enclose in a little house certain rotten flesh, which draws the wild beasts unto it when it stinks. They make a door or a gate of reeds unto the said house through which the filthy smell breaks out and disperses itself into the wide air. The wild beasts take it up and follow it with all the speed they can, for there is not any musk or other sweet things wherewithal men are so much delighted as ravening beasts are with the savor of carrion. Therefore, like an amorous cup, it draws them to the snare of perdition. Besides the rotten flesh, they erect many engines and unavoidable traps to snare in the beast when he comes to raven.

Among the Chaonians, there was a certain young nobleman who loved a virgin called Anthippe, and while the two lovers were walking together a good season in a wood, it happened while they were there that Cichyrus, the King's son, pursued a pardal in hunting, which had fled into that wood. Seeing him, Cichyrus bent his arm against him and cast his dart. The dart missed its mark and killed the virgin Anthippe. The young Prince thought that he had slain the beast and therefore drew near on horseback to rejoice over the fall of the game, according to the manner of hunters. But at his approach he found it far otherwise, for instead of the effusion of the blood of a beast, his right hand had shed the blood of a virgin. When he came to them, he saw her dying and drawing her last breath, and the young man held his hand in the wound to stanch the blood. For sorrow, Cichyrus presently fell distracted in his mind and ran his horse to the top of a sharp rock, from whence he cast down himself headlong and so perished.

The Chaonians, after they had understood this fearful accident and the reason of it, compassed in the place where he fell with a wall, and for the honor of their dead Prince, built a city where he lost his life and called it *Cichyrus* after his own name.

Leopards and panthers do love wine above all other drink, and for this cause Bacchus was resembled to them, and they dedicated to him. All writers do constantly and with one consent affirm that these beasts drink wine unto drunkenness. When the inhabitants of Lybia do observe some little fountain arising out of the land and falling down again and it is known that panthers and pardals are accustomed to drink there early in the morning before it is light, hunters come and pour twenty or thirty pitchers of old sweet wine into the said fountain. Then a little way from it they lie down and cover themselves with clothes or with straw, for there is no shelter either of tree or bushes in that country. In the morning, the panthers, ardently thirsting and being almost dead for want of drink, come unto the fountain, and tasting of the wine, drink thereof great abundance, which falls to work upon their brains, for they begin first of all to leap and sport themselves until they be well wearied, and then they lie down and sleep most soundly, at which time the hunters that lie in wait come and take them without all fear or peril.

Concerning the use of their parts, I find little among the ancients except of their skins. The footmen and ancient soldiers of the Moors did not only wear them for garments but also slept upon them in the nighttime. The shepherds of Ethiopia called *Agriophagoi* do eat the flesh of lions and panthers although it is hot and dry.

If the skin or hide of a leopard being taken and flayed be covered or laid upon the ground, there is such force and virtue in the same that any venomous or poisonous serpents dare not approach into the same place where it is so laid. The flesh of a panther being roasted or boiled at the fire and smelled by anyone who is troubled with palsy or shaking in the joints, or by anyone who is troubled with the beating and continual moving or turning of the heart, is a very profitable and excellent remedy.

To conclude, the brains of a leopard being mingled with a little quantity of the water which is called a canker, and with a little jasmine, and so mixed together and then drunk, mitigate the pain or ache of the belly. The brains of the same beast being mixed with the juice of a canker and anointed upon the genital of any man do incite and stir him up to lechery, but the marrow which comes from this beast being drunk in wine does ease the pain or wringing of the guts and the belly.

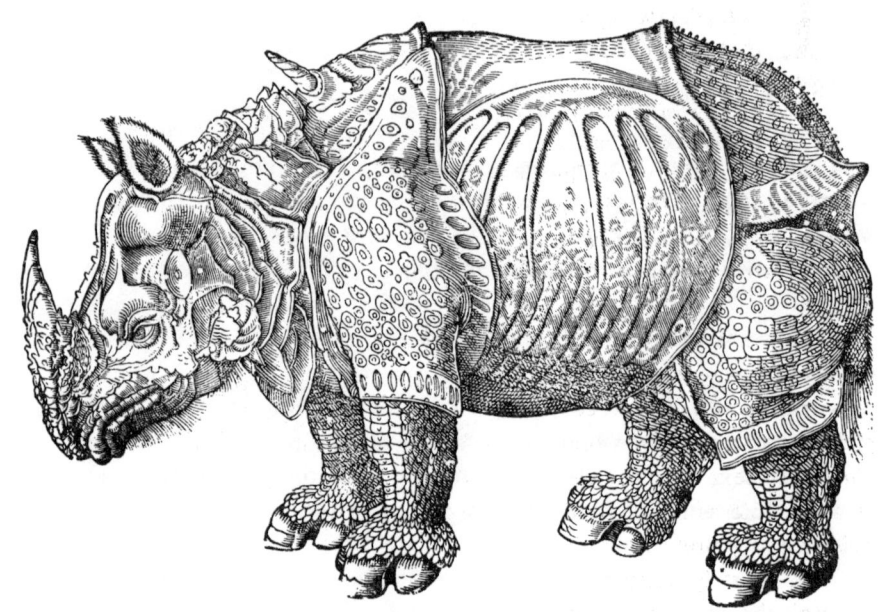

The Rhinoceros

THE RHINOCEROS

TOPSELL SAYS THAT *the rhinoceros is "the second wonder in nature" (the elephant being the first), and he expresses his regret that his account of the animal is not fuller and more interesting than it is.*

The illustration comes from Gesner. Topsell states that Gesner based the illustration on a rhinoceros that was exhibited at Lisbon. According to M. St. Clare Byrne in The Elizabethan Zoo *(London: F. Etchells & H. Macdonald, 1926), Gesner says that he took the illustration from the drawing by Albrecht Dürer. She concludes that Topsell's mistake probably occurred because he misunderstood a section of Gesner's account.*

We are now to discourse of the second wonder in nature: namely, of a beast every way wondrous both for outward shape, quantity and greatness and also for inward courage, disposition and mildness. For, as the elephant was the first wonder of whom we have already discoursed, so this beast next unto the elephant fills up the number, being every way as admirable as he, if he does not exceed him, except in quantity or height of stature.

I am heartily sorry that so strange an outside (yielding no doubt through the omnipotent power of the Creator an answerable inside and infinite testimonies of worthy and memorable virtues comprised in it) should, through the ignorance of men, lie unfolded and obscured before the reader's eyes; for he that shall but see our stories of the apes, the dogs, and small beasts and consider how large a treatise we have collected together out of many writers for the illustration of their natures and conditions, he cannot choose but to expect some rare and strange matters as much unknown to his mind as the outward shape and picture of him appear rare and wondrous to his eyes.

But, gentle reader, you must consider that, since Adam went out of Paradise, there was never any that was able perfectly to describe the universal conditions of all sorts of beasts. For the instruction of man, the Almighty has kept from him the knowledge of many things so that he might learn the difference between his generation and degeneration and consider how great a loss unto him was his fall in Paradise. Before that time man knew God, himself, and the creatures, but since that time he knows neither God as he should know Him, nor himself, nor the creatures as he did know them.

For my part, I write the English story of this beast, and no man should look for anything from my hands which I have not received from some other. I am unwilling to write anything untrue or uncertain out of mine own invention, for truth in every part is so dear unto me that I will not lie to bring any man to love and admire God and His works, for God needs not the lies of men. To conclude this preface, as the beast is strange and never seen in our country, so my eyesight cannot add anything to the description. Therefore, hearken unto that which I have observed out of other writers.

First of all, that there is such a beast in the world, Pliny, Solinus, Diodorus, Aelianus, and others do yield irrefragable testimony. Heliogabalus had one of them at Rome. Pompey the Great in his public spectacles did likewise produce a rhinoceros. When Augustus rode triumphing for Cleopatra, he brought forth to the people a sea horse and a rhinoceros, which was the first time that ever a rhinoceros was seen at Rome. Martial also celebrates an excellent epigram of a rhinoceros which in the presence of Caesar Domitian did cast up a bull into the air with his horn as if he had been a tennis ball. Lastly, to put it out of all question that there is such a beast as the rhinoceros, the picture and figure here expressed was taken by Gesner from the beast alive at Lisbon in Portugal before many witnesses, both merchants and others. So we have the testimony both of antiquity and of the present age for the testimony of the form and fashion of this beast, and that it is not the invention of man but a work of God in nature, first created in the beginning of the world and ever since continued to this present day.

Because of the horn in his nose, the Grecians call him *rhinoceros:* that is, a "nose-horned beast." Although there are many beasts that have but one horn, yet is there none that has one horn growing out of the nose but this beast alone. All the rest have the horn growing out at their foreheads. There have been some people that have taken the rhinoceros for the monoceros (the unicorn) because of this one horn, but they are deceived.

In quantity, the rhinoceros is not much bigger than an oryx. Pliny makes it equal in length to an elephant, and some make it longer than an elephant but say it is lower and has shorter legs. A rhinoceros that was seen at Alexandria had a color like that of an elephant; his quantity was greater than a bull's, or as that of the greatest bull; his outward form and proportion was like a wild boar's, especially in his mouth, except that out of his nose grew a horn, which he used instead of arms. He had two girdles upon his body like the wings of a dragon, coming from his back down to his belly, one toward his neck or mane and the other towards his loins and hinder parts.

To this we may add descriptions out of Oppianus, Pliny, and Solinus. The color of the rhinoceros is like the rind or bark of a box-tree. (This does not differ much from an elephant.) On his forehead there grow hairs which seem a little red, and his back is distinguished with certain purple spots upon a yellow ground. The skin is so firm and hard that no dart is able to pierce it, and upon it appear many divisions like the shells of a tortoise set over the scales, and there is no hair upon the back. Upon his nose there grows a hard and sharp horn, crooking a little towards the crown of his head but not so high. The horn is flat and not round, and it is so sharp and strong that whatever he sets it to, he either casts it up into the air or else bores through it though it be iron or stone.

It is apparent by the picture that there is another horn not upon the nose but upon the withers (I mean the top of his shoulder next to the neck).

Oppianus says that there was never yet any distinction of sexes in rhinoceros, for all that have ever been found have been males and not females. But from hence let nobody gather that there are no females, for it is impossible that the breed should continue without females. Pliny and Solinus say that they engender or admit copulation like elephants, camels, and lions.

When they are to fight, they whet their horn upon a stone. There is not only discord between them and elephants for food, but there is also a natural enmity between the two beasts. It is confidently affirmed that, when the rhinoceros which was at Lisbon was brought into the presence of an elephant, the elephant ran away from him.

How and in what place the rhinoceros overcomes the elephant, we have shown already in the story of the elephant: namely, he fastens his horn in the soft part of the elephant's belly.

All the later physicians do attribute the virtue of the unicorn's horn to that of the rhinoceros, but they are deceived. None of the ancient

Grecians ever observed any medicines in the rhinoceros.

The rhinoceros is taken by the same means that the unicorn is taken, for it is said that above all creatures they love virgins and that unto them they will come, be the beasts ever so wild, and fall asleep before the virgins, and so being asleep, they are easily taken and carried away.

The Tiger

THE TIGER

DURING THE RENAISSANCE, *Europeans knew relatively little about the tiger. To Topsell and his contemporaries, it was a fierce, beautiful animal that was supposed to be incredibly swift. Unlike some writers, Topsell did not believe that all tigers were females and that tigers were generated by copulation with the wind.*

The story that hunters stole tiger cubs by throwing down a glass ball to deceive the mother was a traditional one. In some accounts, the mother is said to see her own image and to believe that it is one of her cubs. In Topsell's account, the mother is said to be deceived by images of the cubs depicted in the ball.

The word *tigris* is an Armenian word which signifies both a swift arrow and a great river. It seems that the name of the river Tigris was so called because of its swiftness, and it seems to be derived from the Hebrew word *gir* and *griera,* signifying a dart.

Tigers are bred in the East, South, and hot countries because their generation desires abundance of heat such as in India and near the Red Sea; and the people called Asangae or Besingi, which dwell beyond the river Ganges, are much troubled with tigers. Likewise, the Prasians, the Hyrcanians, and the Armenians. Apollonius, with his companions traveling between the Hyphasis and the Ganges, saw many tigers. In Berigaza and Dachinabades, which is beyond the Mediterranean region of the East, there is an abundance of tigers and all other wild beasts. In Hispaniola, Ciamba, and Guanassa, there are many lions and tigers.

The Indians say that a tiger is bigger than the greatest horse and that for strength and swiftness tigers excel all other beasts. Some authors say that a tiger fears not an elephant and that one of them has been seen to fly upon the head of an elephant and to devour it. Among the Prasians, when four men were leading a tamed tiger, they met with a mule, and the tiger took the mule by the hind leg, drawing it after him in his teeth, notwithstanding all the force of the mule and the four men. This is unto me a sufficient argument not only of the strength of tigers but of their stature also.

Now, because tigers are strangers in Europe, never breeding in that part of the world and as seldom seen, and because there are not many divers things concerning the nature of the beast, and in the physic, none at all, we must be constrained to make but a short story of it.

The similitude of the body of this beast is like a lioness's, and so are the face and mouth. Their skin is full of spots, not round like a panther's nor divers colored, but square, and sometimes long, and altogether of one color. Yet Solinus and Seneca seem to be of the opinion that their spots are sometimes of divers colors, both yellow and black, and those long like rods. The voice of this beast is called "ranking."

It is needless to speak of their crooked claws, their sharp teeth and divided feet, their long tail, agility of body, and wildness of nature.

It has been falsely believed that all tigers are females and that there are no males among them and that they engender in copulation with the wind. We have already shown in the story of the dog that the Indian dog is engendered of a tiger and a dog, and also the Hyrcanian dogs, whereby it is apparent that tigers do not only conceive among themselves but also in a mingled race. The swiftness of this beast is so great that some people have dreamed that it was conceived by the wind, for as the swiftest horses, and namely the horses of Dardanus, are likewise fabled to be begotten by the Northern wind, so tigers by the West wind.

The male is seldom taken because, at the sight of a man, he runs away and leaves the female alone with her young ones, for he has no care of the whelps, and for this reason I think that the fables first came up that there are no males among tigers.

It is reported that a tame tiger was brought up with a kid and that the said kid was killed and laid before her to eat but she refused it two days together; and the third day, oppressed with extremity of hunger, she made signs by her ranking and crying voice to her keeper for other food. He cast unto her a cat, which she immediately pulled in pieces and devoured.

Generally, the nature of this beast is according to its epithets: sharp, untamed, cruel, ravenous, and never so tamed but it will sometimes return to its former nature. Yet, the Indians do every year give unto their King tamed tigers and panthers, and it comes to pass that sometimes the tiger kisses his keeper.

In the time of their lust, tigers are very raging and furious. The female brings forth many young at once. She nourishes her young in her den very carefully, loving them and defending them like a lioness from hunters, whereby she is many times ensnared and taken.

Only the tiger, the Indians say, can never be conquered, because, when he is hunted, he runs away out of sight as fast as the wind. For this cause, the Indians diligently seek out the caves and dens of the tigers where their young ones are lodged.

Upon swift horses, hunters take and carry away the young ones. When the female tiger returns and finds her den empty, in rage she follows after the hunters, whom she quickly overtakes by reason of her celerity. Seeing her at hand, the hunters cast down one of the whelps. The distressed, angry beast, knowing that she can carry but one, takes up that in her mouth without setting upon the hunters. Contented with that one, she returns with it to her lodging, and, having laid it up safe, back again she returns like the wind to pursue the hunters for the rest. They must likewise set her down another if they have not got into a ship, for, unless the hunters are near the waterside and have a ship ready, she will fetch the young ones from them one by one or else it will cost them their lives. Therefore, that enterprise is undertaken in vain upon the swiftest horses in the world unless the waters come between the hunters and the tiger. The manner of this beast is, when she sees that her young ones are shipped away and she forever deprived of seeing or having them again, she makes so great lamentation upon the seashore howling, braying, and ranking that many times she dies there, but if she recovers all her young ones again from the hunters, she departs with unspeakable joy, without taking any revenge for their offered injury.

For taking away the young ones, hunters do sometimes devise certain round spheres of glass wherein they picture the whelps very apparent to be seen by the mother. One of these spheres they cast down before her at her approach, and when she looks upon it, she is deluded and thinks that her young ones are enclosed therein. Because of the roundness of the sphere, it is apt to roll and stir at every touch. This she drives along backwards to her den, and there breaks it with her feet and nails, and so seeing that she is deceived, she returns again after the hunters for her true whelps. But in the mean season, they are safely harbored in some house, or else they have gone on shipboard.

Johannes Ledesma, a Spaniard, reports this excellent story of a male and female tiger: In the island of Darien, standing in the Occidental Ocean of the new-found world, some eight days sail from Hispaniola, it fell out, in the year 1514, that the said island was annoyed with two tigers, a male and a female, for half a year together so that there was no night free but that they lost a horse, or an ox, or a cow, or a mare, or a hog, or some other animals. It was not safe for men to go abroad in the daytime, much less in the night.

At length, necessity constrained the people to devise a remedy and try some means to mitigate their calamities. Therefore, they sought out all the ways and paths of the tigers to and from their dens so that they might take revenge upon the raveners for the loss of so much blood. At the last, they found a common beaten way. This they cut asunder and dug deep into a large dungeon. Having made the dungeon, they strewed upon the top of it little sticks and leaves, so covering it to dissemble what was underneath. Then came the heedless tiger that way and fell down into the ditch upon such sharp iron stakes and pointed instruments as they had there set. With his roaring he filled all the places thereabout, and the mountain sounded with the echo of his roaring voice.

The people came upon him, and casting great and huge stones upon his back, killed him. But first of all he broke into a thousand pieces the stones, weapons, and spears that were cast against him; and so great was his fury that, when he was half dead and the blood ran out of his body, he terrified the persons looking upon him.

The male tiger being thus killed, the people followed the footsteps into the mountains where the female was lodged, and there in her absence took away two of her young ones, but afterward changed their minds and carried them back again, putting upon them two brazen collars and chains and making them fast in the den so that, when the young ones had sucked till they were greater, they might be with pleasure and safety conveyed into Spain. At last, when the time had come that they should be taken forth to be sent away, the people went to the den, and therein they found neither young nor old but the collars fastened in the same place that they had left them. Thereby it was conceived that the envious mother had killed and torn her young ones in pieces rather than they should fall into the hands of the hunters. So this love of hers ended in horrible cruelty.

In ancient times tigers were dedicated to Bacchus, as all spotted beasts were, and tigers did draw his chariot while he did hold the reins.

Notwithstanding their great minds and untamable wildness, tigers have been taken and brought in public spectacle by men, and the first of all that ever brought them to Rome was Augustus when Quintus Tubero and Fabius Maximus were consuls, at the dedication of the

Theater of Marcellus, the which tigers were sent unto him out of India for presents. Afterwards, Claudius presented four to the people, and lastly Heliogabalus caused his chariots to be drawn with tigers.

The Indians near the river Ganges have a certain herb growing like bugloss. They take it and press the juice out of it, and in still, silent nights, they pour it down at the mouth of the tigers' den, by virtue whereof it is said the tigers are continually enclosed, not daring to come out over it because of some secret opposition in nature, and famish and die, howling in their caves through intolerable hunger.

It is reported that, when tigers hear the sound of bells and timbrels, they tear their own flesh from their backs.

Ledesma, of whom we spoke before, affirms that he did eat of the flesh of the tiger that was taken in the island of Darien, and that the flesh was nothing inferior to the flesh of an ox. But the Indians are forbidden by the laws of their country to eat any part of a tiger's flesh except the haunches.

The Unicorn

THE UNICORN

To most modern readers, the word unicorn calls to mind a creature like the one seen in the illustration (a creature with a horse's body). Topsell also describes other types of unicorns, but in doing so he was following a position taken by many writers of his time. Topsell makes clear that he is talking about the "true" unicorn, not another one-horned creature. He believed that doubting the existence of the unicorn involved a lack of faith in the power and glory of God.

A large part of the history deals with the medicinal virtues connected with the unicorn's horn. Topsell was skeptical of some of the virtues attributed to the horn; but, like most learned men of the period, he believed that the horn had many marvelous powers.

We are now come to the history of a beast of which divers people in every age of the world have made great question, because of its rare virtues; therefore, it behooves us to use some diligence in comparing together the several testimonies that are given of this beast. The main question to be resolved is whether there be a unicorn.

To begin with, by the unicorn we do understand a particular beast which has naturally but one horn, and that a very rich one, that grows out of the middle of the forehead. There are divers beasts that have but one horn. But our discourse of the unicorn is of none of these beasts, for there is not any virtue attributed to their horns. Our discourse is of the true unicorn, and because of the nobleness of his horn, he has ever been in doubt.

That there is such a beast Scripture itself witnesses, for David thus speaks in Psalm 92: "My horn shall be lifted up like the horn of a unicorn." All divines that have ever written have not only concluded that there is a unicorn, but also affirm the similitude between the kingdom of David and the horn of the unicorn, for as the horn of the unicorn is wholesome to all beasts and creatures, so should the kingdom of David be in the generation of Christ. Do we think that David would compare the virtue of his kingdom and the powerful redemption of the world unto a thing that is not or is uncertain and fantastical?

Likewise, in many other places of Scripture, we will have to traduce God, Himself, if there is no unicorn in the world.

All our European authors that write of beasts make of unicorns divers kinds. Many beasts have not only their divisions, but their subdivisions into subalternal kinds, as many dogs, many deer, many horses, many panthers. Why should there not also be many kinds of unicorns? If the reader is not pleased with this, let him show me better reason or else be silent, lest the uttering of his dislike reveal envy and ignorance.

Pliny says that the Arcean Indians hunt a certain wild beast which is untamable and has one horn. In the head, this beast resembles a hart; in the feet, it is like an elephant; in the tail, like a boar; and in the rest of its body, like a horse. Its horn is about two cubits long. Aelianus says that there are certain mountains in the midst of India where are abundance of wild beasts, and among other beasts there are unicorns. In their ripe old age, they are as big as horses; their mane and hair are yellow. They excel in the celerity of their feet and bodies and have feet cloven like an elephant's, the tail of a boar, and one black horn growing out between their eyebrows. This horn is not smooth, but rough all over with wrinkles. It grows to a most sharp point.

It appears that both Pliny and Aelianus are describing the same beast. I gather that this could be the wild ass of India, for the wild ass is about the size of a horse and has a horn in the middle of its forehead, and it is a swift beast of solitary life and of exceeding strength and an untamable nature. But the wild ass is white in the body and purple on the head, and Aelianus says that its horn also differs in color from the horn of the unicorn.

Therefore by what is said it appears to me that either the Indian ass is a unicorn, or differs from it only in color.

Ludovicus Roman saw two unicorns at Mecca in Arabia where Mahomet's temple and sepulcher is. They are preserved within the walls and cloisters of that temple. Now, their description is of this sort: one of them and the elder was about the stature of a colt of two years and a half old, having a horn growing out of his forehead of two cubits length, and the other unicorn was much smaller, for it was but a year old and like a colt of that age; its horn was some four spans long or thereabouts. The head of them was like the head of a hart, the neck not long and the mane growing all on one side. The legs were slender and lean, like the legs of a hind, and the hoofs of the forefeet were cloven like the feet of a goat; the hind legs were all hairy and shaggy on the outside. Although these beasts were wild, yet by art or superstition, they seemed to be tempered with no great wildness.

It is said that, in the kingdom of Basman, which is subject to the great Cham, there are unicorns somewhat lesser than elephants, having hair like oxen, heads like boars, feet like elephants, and one horn in the middle of their foreheads. I deem this beast to be a second kind of unicorn.

It is also reported that, in Masinum or Serica (that is, the mountains between India and Cathay), there is a certain beast having a swine's head, the tail of an ox, the body of an elephant, and one horn in the forehead. Now, if the reader shall think this beast different from the former, I make it the third kind of unicorn.

Aloisius Cadamustus writes that there is a region of the new-found world wherein are found live unicorns; and toward the East and South, under the Equinoctial there is a living creature (with one horn which is crooked and not great) having the head of a dragon, and a beard upon the chin, the neck long and stretched out like a serpent's; the rest of the body is like to a hart's, except that the feet, color, and mouth are like a lion's. This beast (if not a fable or rather a monster) may be a fourth kind of unicorn.

Unicorns are very swift. They keep for the most part in the deserts and live solitary in the tops of mountains. There is nothing more horrible than the voice or braying of the unicorn, for his voice is strained above measure.

The unicorn fights with both the mouth and his heels, with the mouth biting like a lion's and with the heels kicking like a horse's. He is a beast of an untamable nature. He fears not iron nor any iron instrument, and what is most strange of all other is that he fights with his own kind (yea, even with females unto death, except when he burns in lust for procreation), but unto stranger-beasts, with whom he has no affinity in nature, he is more sociable and familiar, delighting in their company when they come willingly unto him, never rising against them, but proud of their dependence and retinue, keeps with them all quarters of league and truce. With his female, when once his flesh is tickled with lust, he grows tame, gregarious, and loving, and so continues till she is filled and great with young, and then returns to his former hostility.

The unicorn is an enemy to the lion, wherefore, as soon as ever a lion sees a unicorn, he runs to a tree for succor, so that, when the unicorn makes force at him, he may not only avoid his horn but also destroy the unicorn, for, in the swiftness of his course, the unicorn runs against the tree wherein his sharp horn sticks fast. Then, when the lion sees the unicorn fastened by his horn, he falls upon him and kills him.

It is said that unicorns above all other creatures do reverence virgins and young maids, and that many times at the sight of them, unicorns grow tame, and come and sleep beside them, for there is in their nature a certain savor by which the unicorns are allured and delighted. The Indian and Aethiopian hunters are said to use a stratagem to take the beast. They take a goodly strong and beautiful young man, whom they dress in the apparel of a woman, besetting him with divers odoriferous flowers and spices. The man so adorned, they set him in the mountains or the woods where the unicorn hunts, so as the wind may carry the savor to the beast, and in the mean season, the other hunters hide themselves. Deceived by the outward shape of a woman and the sweet smells, the unicorn comes unto the young man without fear and so suffers his head to be covered and wrapped within his large sleeves, never stirring but lying still and asleep, as in his most acceptable repose. Then when the hunters by the sign of the young man perceive the unicorn fast and secure, they come upon him and by force cut off his horn and send him away alive.

Concerning this opinion we have no elder authority than Tzetzes, who did not live above five hundred years ago, and therefore I leave the reader to the freedom of his own judgment to believe or refute this relation.

We will next relate the true history of the horn of the unicorn. The horn grows out of the forehead between the eyelids. It is neither light nor hollow, nor yet smooth like other horns, but hard as iron, rough as a file. It is wreathed about with divers spires. It is sharper than any dart, and it is straight and not crooked, and everywhere black except at the point.

The horn of the unicorn has a wonderful power of dissolving and expelling all venom or poison. If the unicorn puts his horn into water from which any venomous beast has drunk, the horn drives away poison, so that the unicorn can drink without harm. It is said that the horn being put upon the tables of kings and set among their junkets and banquets reveals any venom if there be any such therein, by a certain sweat which comes over the horn. But that it sweats thus is false; it does perhaps sometimes sweat, even as some solid, hard, or light substance might do if some external vapor is about it; but this does not appertain to poison.

THE UNICORN

To conclude, I will discuss the remedies which writers have attributed to the horn of the unicorn. Rich men do usually cast little pieces of the horn in their drinking cups, either for preventing or curing some disease. There are also some people who enclose it in gold or silver, and so cast it in their drink, as though the force of it could remain many years, notwithstanding the continual soaking in wine. Most men, for the remedies arising from the horn, use it simply by itself. Others prefer the marrow therein.

The horns of unicorns, especially that which is brought from new islands, being beaten and drunk in water, help wonderfully against poison. I have myself heard of a man worthy to be believed that, having eaten a poisoned cherry and perceiving his belly to swell, he cured himself by the marrow of the horn being drunk in wine.

We drink the substance of the horn either by itself or with other medicines. Once, I happily made this Sugar of the horn, as they call it, mingling with it amber, ivory dust, leaves of gold, coral, and certain other things, the horn being included in silk and beaten in the decoction of raisins and cinnamon. I cast them in water. In the meantime I did not neglect other ways to heal myself.

The horn of a unicorn being beaten and boiled in wine has a wonderful effect in making the teeth white or clear. And thus much shall suffice for the medicines and virtues arising from the unicorn.

The Wolf

THE WOLF

Topsell describes the customary attributes associated with the wolf: treachery, deceit, hypocrisy, ravenousness, and cruelty. He gives many of the strange beliefs that were current about wolves and includes an interesting description of lycanthropy. He was not among those writers who believed that people could turn into wolves.

There are divers kinds of wolves in the world. Oppianus in his admonition to shepherds makes mention of five kinds.

The first kind is a swift wolf and runs fast and is called therefore *toxeuter,* that is, *sagittarius,* a "shooter." This kind has a greater head than other wolves and likewise greater legs fitted to run, white spots on the belly, round members, and his color is between red and yellow. He is very bold, howling fearfully, having fiery, flaming eyes, and continually wagging his head.

The second kind has a greater and larger body than the first, being swifter than all other wolves. These wolves are called *harpages,* and they are the greatest raveners, to whom our Savior Christ in the Gospel compares false prophets when he says, "Take heed of false prophets which come unto you in sheep's clothing, but are inwardly *lyco harpages,* ravening wolves." The sides and tail of this kind of wolf are of a silver color. He lives in the mountains, except in the wintertime, wherein he descends to the gates of cities or towns and boldly without fear kills both goats and sheep, yet by stealth and secretly.

The third kind is called *lupus aureus,* a golden wolf, by reason of his color. For beauty, he is worthily preferred before the others because of his golden, resplendent hairs, and therefore Oppianus says that he is not a wolf but some wild beast excelling a wolf. He inhabits the white rocks of Taurus and Cilicia or the tops of the hill Amanus and such other sharp and inaccessible places. He is exceedingly strong, especially being able with his mouth and teeth to bite asunder not only stones but brass and iron. He fears the Dog Star and heat of summer, rejoicing more in cold than in warm weather. Therefore, in the dog days, he hides himself in some pit or opening of the earth until that sunny heat is abated.

The fourth and fifth kinds are called by one common name *acmonae*. Now *acmon* signifies an eagle or else an instrument with a short neck, and it may be that these are so called in resemblance of the ravening eagle, or else because their bodies are like to that instrument, for they have short necks, broad shoulders, rough legs and feet, small snouts, and little eyes. The one kind has a back of a silver color and a white belly, and the lower part of the feet is black, and this kind is called *ictinus,* a gray kite-wolf. The other kind is black, having a lesser body, his hair standing continually upright, and he lives by hunting of hares.

There are some authors who have thought that dogs and wolves are one kind: namely, that vulgar dogs are tame wolves and that ravening wolves are wild dogs. But Scaliger has learnedly confuted this opinion, showing that they are two distinct kinds.

Although there is a great difference of colors in wolves, yet most commonly they are gray and hoary, that is, white mixed with other colors. The brains of a wolf do decrease and increase with the moon, and their eyes are yellow, black, and very bright, sending forth beams like fire, and carrying in them apparent tokens of wrath and malice; and, for this cause, it is said that they see better in the night than in the day. In ancient times, the wolf was dedicated to the sun, for the quickness of his seeing sense and because he sees far. Also his sense of smell is of great quickness. It is reported that, in time of hunger, he smells his prey a mile and a half or two miles off by the benefit of the wind.

Their teeth are smooth, sharp, and unequal and therefore bite deep. All beasts that are devourers of flesh do open their mouths wide that they may bite more strongly, and especially the wolf.

The neck of a wolf stands on a straight bone that cannot well bend. Therefore, like the hyena, the wolf must turn round about when he would look backwards. The neck is short, which argues a treacherous nature.

It is said that wolves will swim and go into the water two by two, every one hanging upon another's tail, which they take in their mouths. Therefore they are compared to the days of the year, which do successively follow one another. By this successive swimming, they are better strengthened against impression of the floods and are not lost in the waters by any overflowing waves or billows.

Great is the voracity of these beasts. They are so insatiable that they devour hair and bones with the flesh which they eat. When they are hungry, they rage much. Although they might be nourished tame, yet can they not abide any man to look upon them while they eat. When wolves have driven away their hunger with abundance of meat, they are unto men and beasts as meek as lambs. Till they be hungry again, they are not moved to rapine, though they go through a flock of sheep. But, in short time after, their bellies and tongue are calling for more meat, and they return to their former conditions and become as ravening as before.

It is said that wolves do eat a kind of earth called *argilla,* which they do not eat for hunger but to make their bellies heavy, to the intent that, when they set upon a horse, an ox, a hart, an elk, or some such strong beast, they may weigh the heavier and hang fast at their throats till they have pulled them down. By the virtue of this tenacious earth, their teeth are sharpened and the weight of their bodies increased. When they have killed the beast that they have set upon, they disgorge themselves and empty their bellies of the earth as unprofitable food by a kind of natural vomit, before they touch any part of the flesh of the beast.

Wolves have this in common with lions: in their greatest extremity of hunger, when they have the choice of a man or a beast, they forsake the man and take the beast.

Some people are of the opinion that, when wolves are old, they grow weary of their lives and therefore come unto cities and villages, offering themselves to be killed by men. But this is a fable.

The wolves that are most sluggish and least given to hunting are most ready to venture upon men, because they love not to take much pains in getting their living. It is reported that a wolf will never venture upon a living man unless he has formerly tasted of the flesh of a dead man. But of these things I have no certainty but rather do believe the contrary.

Wolves are enemies to asses, bulls, and foxes, for they feed upon their flesh, and there is no beast that they take more easily than as ass. They also devour goats and swine of all sorts, except boars, which do not easily yield unto wolves. It is said that, when they will deceive goats, they come unto them with the green leaves and small boughs of osiers in their mouths, wherewithal they know goats are delighted, so that they may draw them therewith as to a bait to devour them.

When a wolf falls upon a goat or a hog or some such other beast of small stature, his manner is not to kill it but to lead it by the ear with all the speed he can drive it to his fellow wolves; and if the beast is stubborn and will not run with him, he beats its hinder parts with his tail. He causes the poor beast to run as fast or faster than himself unto the place of its own execution, where it finds a crew of ravening wolves to entertain it. At its first appearance, they seize upon it and, like devils, tear it to pieces in a moment, leaving nothing uneaten but its bowels.

Generally, all wolves are enemies to sheep. The foolish sheep in the daytime are easily beguiled by the wolf. At the sight of the sheep, the wolf makes an extraordinary noise with his foot, whereby he calls the foolish sheep unto him. Standing amazed at the noise, a sheep falls into his mouth and is devoured. When the wolf comes to the fold of sheep in the nighttime, he first of all compasses it round about, watching to find out whether the shepherd and the dog are asleep or awake. If they are present and like to resist, then he departs without doing any harm, but if they are absent or asleep, then loses he no opportunity but enters into the fold and falls a killing, never giving over till he has destroyed all, unless he is hindered by the approach of someone. His manner is not to eat any till he has killed all, not because he fears that the over-livers will tell tales, but because his insatiable mind thinks he can never be satisfied. When all are slain, he eats one of them.

When wolves chance to see an ox in the water or in a marsh encumbered with mire, they come round about him, stopping all the passages where he should come out, baying at him and threatening him so as the poor, distressed ox plunges himself many times over head and ears, or at the leastwise they so vex him in the mire that they never suffer him to come out alive. At last, when they perceive him to be dead and clean without life by suffocation, it is notable to observe their singular subtlety to draw him out of the mire, whereby they may eat him. One of them goes in and takes the beast by the tail and draws him with all the power he can. But wit without strength may better kill a live beast than remove a dead one out of the mire. Therefore, he looks behind him and calls for more help. Then another wolf takes the tail of the first wolf in his mouth; and a third wolf takes the tail of the second; a fourth, the tail of the third; and so on, increasing their strength until they have pulled the beast onto the dry land.

It is reported that in time of great famine, when wolves get no meat, they destroy one another, for when they meet together, they run round in a circle as it were by consent, and that wolf which is first giddy, being not able to stand, falls down to the ground and is devoured by the rest, for they tear him in pieces before he can arise again.

Some writers say that, if a wolf first sees a man, the man becomes silent and cannot speak, but if the man sees the wolf first, the wolf is silent and cannot cry. Although these things are reported by Plato and some other authors, I rather believe them to be fabulous than true.

It is reported that, if a mare with foal treads upon the footsteps of a wolf, she casts her foal. The Egyptians, when they signify abortment, do picture a mare treading upon a wolf's foot.

Wolves were sacred to Apollo, Jupiter, and Mars, and we read of Apollo Lycius or Lyceus, to whom there were many temples built, and of Jupiter Lyceus, the sacrifices instituted unto him called Lycoea, and games by the same name. There were other holy days called Lupercalia, wherein barren women did chastise themselves naked because they could bear no children, hoping thereby to gain fruitfulness of the womb.

Wolves fear fire even as lions do, and therefore persons who travel in woods and secret places by night, wherein there is any suspicion of meeting wolves, carry with them a couple of flints wherewithal they strike fire at the approach of the ravening beast, and this so dazzles his eyes and daunts his courage that he runs away fearfully. It is said that wolves are afraid of the noise of swords or iron struck together. A man traveling near Basel, with a bell in his hand, threw stones at a wolf that followed him, and it nothing availed. When he by chance fell down and the bell did give a sound, the wolf being affrighted ran away. When the man perceived this, he sounded the bell aloud, and so drove away the wild, ravening beast.

If at any time a wolf follows a man afar off, as it were treacherously to set upon him suddenly and destroy him, let him but set up a stick or staff, or some such other knowledgeable mark, in the middle space between him and the wolf, and it will scare him away, for the suspicious beast fears such a man and thinks that he carries about him some engine or trap to take away his life. It is also said that, if a traveler draws after him a long rod or pole or a bundle of sticks and clouts, a wolf will never set upon him, mistrusting some policy to overthrow and catch him. There are some persons that take wolves by poisoning, for they poison certain pieces of meat and cast them abroad, whereof, when the wolves eat, they die immediately.

There is a disease called the *wolf* because it consumes and eats up the flesh in the body next to the sore, and it must every day be fed with fresh meat, as lambs, pigeons, and such other things wherein is blood, or else it consumes all the flesh of the body, leaving not so much as the skin to cover the bones.

There is a certain territory in Ireland where the inhabitants are foolishly reported to be turned into wolves when they are past fifty years old. The reason for this belief is conjectured to be that, for the most part, the inhabitants are vexed with the disease called *lycanthropy,* which is a kind of melancholy causing the persons so affected about the month of February to forsake their own dwelling or houses and to run out into the woods, or near the graves and sepulchers of men, howling and barking like dogs and wolves. The true signs of this disease are thus described by Marcellus. Those who are affected have their faces pale and their eyes dry and hollow, looking drowsy; and they cannot see; their tongues are very rough and look as though they were all scabbed. Neither can they spit, and they are very thirsty, having many ulcers breaking out of their bodies, especially on their legs.

Some call this disease *Lycaon,* and men oppressed therewith *Lycaones* because it has been feigned by the poets that there was one Lycaon, who for his wickedness or his sacrificing of a child was by Jupiter turned into a wolf, being utterly distracted of human understanding.

Wolves are enemies to all, and they take special revenge against them that harm them.

Some writers say that, when many wolves have obtained a prey, they do equally divide it among them all, but whether this is true I cannot tell. Rather I think the contrary, because they are insatiable and never think they have enough.

When wolves set upon sheep, they choose a dark, cloudy day or time so that they may escape more freely; and to the intent that their treadings should not be heard, they lick the bottom or soles of their feet, for by that means they make no noise among the dry leaves. And, if going along, a wolf chances to break a stick and so against his mind makes a noise, he bites his foot as if it were guilty of that offense.

For the most part, wolves set upon animals that have no keepers and raven in secret. If they come unto a flock of sheep where there are dogs, they first of all consider whether they are able to make their party good, for, if they see that they cannot match the dogs, they depart away although they have begun their spoil. But, if they perceive their forces to be equal or superior, then they divide themselves into three ranks.

One company of them kills sheep, a second company fights with the dogs, and the third sets upon the men.

When wolves are in danger of being taken by hunters, they bite off the tip of their tails. Therefore, the Egyptians, when they would describe a man delivered out of extremity and danger, do picture a wolf lacking that part of his tail.

When wolves are in peril, they are extremely fearful, astonished, and afraid. When they are shut up, they seem harmless. This argues the baseness of their mind, which is subtle, cowardly, and treacherous, daring to do nothing but for the belly and not then but upon a singular advantage.

Once, a certain wolf, driven by hunger, came into a village near Milan in Italy and there entered into a certain house wherein sat the good wife and her children. Being terrified and not knowing what she did, the poor woman ran out of the house, pulling the door to after her and so shutting the wolf in among her children. At last, her husband returned home, unto whom she related the accident and how she had shut up the wolf. The man entered hastily indoors, longing to save and deliver his poor infants. When he came in, he found all well. The wolf was in worse case, for it stood like a stock without sense, astonished, amazed, and daunted. It was not able to run away and stood as if offering itself to be destroyed.

The great-uncle of Goblerus, being marvelously addicted to the hunting of wild beasts, had in his land divers ditches and pits and trenches cast up for the safekeeping of such beasts as should fall into them. Now, it happened that, upon one Sabbath at night, there fell into one of those pits three creatures of divers disposition and adverse inclination and that none of them was able to get out.

The first was a neighbor's wife, a poor woman, who, going to the field to gather beets and rapes for the following day, happened by mischance to fall down into the pit. There she was forced to lodge all night (you must think with great anguish, sorrow, and perilous danger to herself). She had not tarried long in the pit ere a fox was likewise taken and fell down upon her. Now began her grief to be increased. Having no means to escape, she feared that the wild beast would bite and wound her. She cried as loud as ever she could, and after she had wearied herself, necessity made her to be patient, being a little comforted to see that the fox was as much afraid of her as she was of him. Yet, she thought the night full long, wishing for the break of day when men would stir abroad to their labors, hoping that someone would hear her lamentation and deliver her from the society of such a chamber fellow.

While she thus thought, striving between hope, fear, and grief, suddenly a wolf was taken and fell down upon her. Then she lost her hope, and in lamentable manner thinking of husband and children, how little they conceived of her extremity, she resolved to forsake the world and commended her soul to God, making no other reckoning but that her distressed, lean limbs should now be a supper and breakfast to the wolf, wishing that she might but see her husband and kiss her children before she lost her life by that savage execution. While she thus mused, she saw the wolf lie down, she sitting in the one corner and the fox resting in another, and the wolf appalled as much as either of them. So the woman had no harm but an ill night's lodging, with the fear whereof she was almost out of her wits.

Early in the morning, the great-uncle of Goblerus came to see what was taken in his trenches and pits, and coming unto that pit, he found a treble prey: a woman, a wolf, and a fox. He was greatly amazed and stepped a little backward. Seeing him, the woman cried out, calling him by his name and praying for his aid. Knowing her by her voice, he immediately leaped down into the pit (for he was a valiant man) and with his weapon first slew the wolf and then the fox, and so delivered the woman from the fear of them. Yet, he was forced to leave her there till he went and fetched a ladder, for she was not able to come forth as he was. Having brought the ladder, he went down again into the pit and brought her forth upon his shoulders, in that manner delivering her safe to her husband and family.

Now these two stories do plainly set forth that a wolf dares do nothing when he is in fear himself.

It is a question whether wolves can be tamed or not. Some say that they are always wild and can never be tamed. Albertus writes that being taken as whelps they can be tamed and will play like dogs. Yet, he says that they never forget their hatred against the hunter and that, when they go abroad, they never forget their desire for lambs or other beasts which are devoured by wolves.

Wolves are subject to the same diseases that dogs are, especially the swellings of the throat, madness, and the gout. The reason why dogs and wolves are more subject to madness than any other beast is that their bodies are choleric and that their brains increase and decrease with the moon. If a man is bitten by a mad wolf, he is to be cured by the same medicines that are applied to the bitings of a mad dog.

Wolves live very long, even until they lose their teeth, and when they are oppressed with famine in their old age, they fly unto cities and houses to seek food.

A wolf was once part of the arms of Rome, and the judgment seat at Athens had in it the picture of a wolf. There were ancient coins of money stamped with the image of a wolf, both among the Grecians and among the Romans, which were therefore devised because Romulus and Remus were said to be nursed by a wolf.

There is a four-footed beast called the sea-wolf. This beast lives both on sea and land, satisfying its hunger for the most part upon fishes. It resembles the wolf that lives on the land. It has very many hairs growing on both sides of its eyes to shadow them. Its nostrils and teeth are like a dog's, and there are strong hairs growing about its mouth. There are also small bristles growing upright upon its back, and it is adorned and marked on each side with black, distinct spots. It has a long tail, thick and hairy. In all other parts, it is like a wolf.

The blood of a wolf being mixed with oil is very profitable against deafness of the ears. The dung and blood of a wolf are much commended for those that are troubled with the colic and stone. A wolf's flesh being boiled and taken in meat helps those that are lunatic. A wolf's flesh being eaten is good for procreation of children.

Being anointed upon those whose joints are broken, the fat of a wolf does very much profit. Some of the later writers were wont to mingle the fat of the wolf with other ointments for the disease of the gout. Some also do mingle it with other ointments for the palsy. The head of a wolf, being burned into ashes, is a special remedy for the looseness of the teeth. A wolf's right eye, being salted and bound to the body, drives away all agues and fevers. The right eye is very good against the bitings of dogs.

The teeth of a wolf, being rubbed upon the gums of young infants, do open them, whereby the teeth may the easier come forth. The heart of a wolf, being burned and beaten to powder and so taken in drink, helps those that are sick of the falling sickness. And thus, with this recital of the medicines arising from the wolf, we bid him farewell.

AN EPILOGUE TO THE READERS

I TAKE MY OWN conscience to witness that in this work I have intended nothing but His glory that is the creator of all; and if I thought that hereby the readers would not be the more moved to acknowledge and obey His sovereign majesty, but that they read these stories to feed curiosity, I would not only desist and go no farther but wish that this work were buried in oblivion and that the poor paralytic hand which wrote and indited it were severed from the body. Therefore (well-minded readers), herein you shall satisfy your own consciences and hearts, when the visible things of the world do lead you to the invisible things of God and when all these rows and ranks of beasts are as letters and midwives to save the reverence which is due to the Highest from perishing within you.

If you think my endeavors and the printer's costs necessary and commendable and if you would ever further or second a good enterprise, I do require all men of conscience that shall ever hear, read, or see these histories, or wish for the sight of the rest, to help us with knowledge and to certify their particular experiences in any kind or any one of the living beasts.

I conclude with the saying of St. Augustine in his Book of Gen. against the Manichees, where he speaks thus of the beasts and all creatures: "Make use of the profitable things, beware of the pernicious, relinquish the superfluous; nevertheless, in all of them, when you see proportion and harmony in order, look for the Creator."

Farewell